GRAYS HARBOR COLLEGE
The JOHN SPELLMAN LIBRARY
1620 EDWARD P. SMITH DR.
ABERDEEN, WA 98520-7599

DISCARDED
By Grays Harbor College Library

AF681472

THE HEALING SEA

Drawings by Stavros Georgeulakos

The

A VOYAGE INTO THE ALIEN

Dr. George D. Ruggieri, S.J.
Norman David Rosenberg

Healing Sea

WORLD OFFSHORE

DODD, MEAD & COMPANY, NEW YORK

49104

Portions of this material have appeared in different form in *Oceans*, *Surveyor*, *Pegasus*, *Science*, and the *Annals of the N.Y. Academy of Sciences*.

No attempt has been made by the authors to catalogue every marine substance of biomedical interest.

Copyright © 1969, 1970, 1974, 1975, 1978
by George D. Ruggieri, S. J., and Norman David Rosenberg
All rights reserved
No part of this book may be reproduced in any form
without permission in writing from the publisher
Printed in the United States of America

1 2 3 4 5 6 7 8 9 10

Library of Congress Cataloging in Publication Data

Ruggieri, George D.
The healing sea.

Includes index.
1. Marine biology. I. Rosenberg, Norman David, joint author. II. Title.
QH91.15.R83 574.92 78-7333
ISBN 0-396-07449-9

THE LIBRARY
GRAYS HARBOR COLLEGE
ABERDEEN, WASH. 98520

FOR ROSS AND BELLE AND IRVING

CONTENTS

	Prologue: Dawn	1
I.	Morning	7
II.	Afternoon	55
III.	Night	103
	Epilogue	149
	Glossary	161
	Index	165

PROLOGUE

Dawn

After the Core too old for imagining exploded
and the congealing dust united as Sol and Earth,
the emerging land, molten at heart and quaking now,
was embroiled in storms of such shattering violence
that had the weakest one beaten upon our eyes it
would have struck home as the absolute end of things;
the final blow against all mortal life. Overhead
the cocooning air seethed with turbulent poisons,
a mixture of heavy gases and hydrous vapors seeping

from the fractured crust and erupting in looming gusts that swallowed light, paling the sun, each time the infant earth belched out another burst of liquid rock, mounting upon itself the shape of continents to come. Nothing was there to feed upon the hissing clouds as they were taking cover beyond the cauldron ground; by gravity's lengthening arm denied full flight back into the quieter well of Space where the free-floating hydrogen had gone, leaving behind an encircling stampede of methane and ammonia that was giving the sky a boundary, making the heavens roar.

Before the sea had risen and nursed the first dividing cell to completion in its embryonic flow, the solitary pieces of Existence were seeded in the wind like summer pollen; organic molecules coaxed from gas by lightning, airborne as a sudden, deeper flashing that fell into the soil bed, waiting to be seized by another founding spark. Bathed in ultraviolet light, the original air was unfolding those substances essential to elemental life: amino and nucleic acids, proteins and carbohydrates, the cornerstones that would somehow be combined in limpid, wriggling, plasmic shapes, each with a compulsion for being continued, chosen to race in the maze of faces that leads from bare bacteria to humankind.

For millions of years lightning clawed the smoking bubble of expanding atmosphere, like diamond streaks cut in glass and just as quickly healed; while below, mountains of steam reared up from the lava and became dark herds of boiling thunderheads that cooled aloft and then poured down an agelong avalanche of rain. Descending from frayed, condensing clouds like blinding silver beams, fresh water scored the open ground, in every scoop and hollow gouging streams that spilled again from high to low, seeking a bottom for gathering the initial disjointed flow of river and creek as puddle grown, claiming new height by adding an ocean of depth below. Now one engulfing body, the Paleo-Pacific and its amniotic seas submerged the planet, scaling every grain of land except for a single thrusting continent that massed itself from pole to pole, holding solid surface to the air. But like the harder ocean basin rock, it rested on divided, shifting crust; under pressure, slowly it was casting off its parts.

Scissored by tension cracks that widened into lava-spouting rifts, Pangea departed as a fleet of continents carried by the planet's inner tides on crustal plates that sever and collide, leaving in their molten wake another valley stretching into ocean, or forcing by their impact the disappearance of a cornered sea in wreckage that becomes a mountain rise. One rupturing seam realigned the motherland's eastern coast, laying in between Africa and the Americas the ever-spreading Atlantic floor. Other fissures pushed India to sail north from Antarctica's tearing shore until this drifting apex of the next bestirring ocean beached upon Eurasia in a monumental crash, plowing for its headstone the titanic Himalayas.

Trickling past the tidal flats, slipping from the shore, streams of mud-soaked active molecules found homefall in the sea; thus enjoined began to spawn. In a garden of surface froth and foam, they added process to each other, came alive; bubbling mouths that nibbled on sun and gas, manufacturing the power to survive. In microscopic mystery, an urge for replication, the unending had been formed; away from the unwet earth that petrifies all chance, grinds down its quiet compounds like lifeless flecks of sand. Curled in water breath was blown, cells began a dance. More perfectly devised than any later drawn for drier skies is the experiment nature held inside, cushioned in the molding fluid; track of the aquabound.

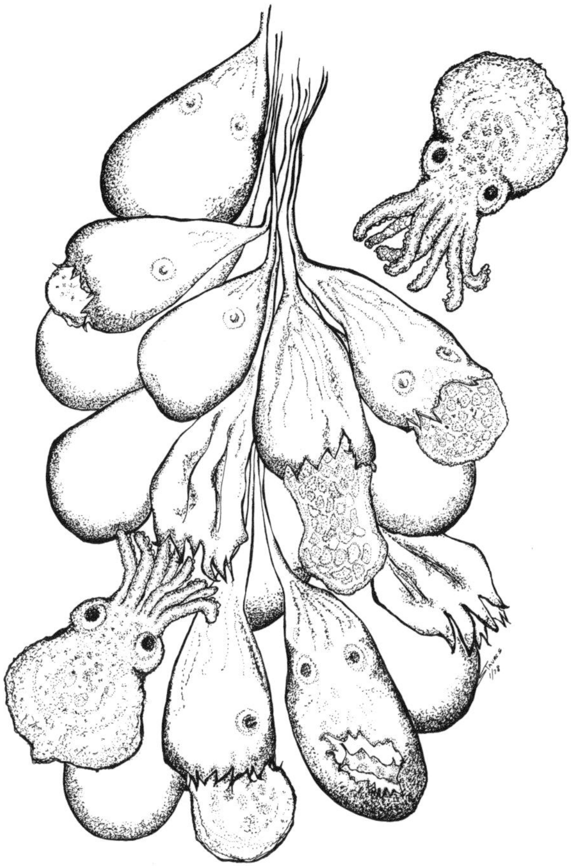

I. Morning

Octopus babies hatching

In the warm shallows fringing an outlying island, a solitary coral larva, so tiny it is barely visible, is sinking to the ocean floor. For some days it has been drifting through the golden waters of this pacific sea, urged on by even the gentlest currents and the fluttering strokes of the cilia covering its pear-shaped body. But all the while it has been growing heavier, and now its added weight is inexorably driving it down—one more drop in the endless rain of minute particles sprinkling over the ocean bottom. There its survival will first depend upon the accidental location of its landing—the blind coincidence of an unguided fall ending on a stone or a shell or some other hard substance which has not been fouled by mud or clumps of algae.

Settled by chance on the clean, firm surface of a recently sunken ship, the larva immediately begins secreting a limestone home for itself with the calcium carbonate it has drawn from the water. Permanently anchored inside this calcareous cup, it can then complete its transformation into a delicate polyp—a thin tube of translucent tissue crowned by a mouth which is rimmed with petallike tentacles. Within the polyp's shivery tissue, countless phytoplankton swarm, enhancing its fragile color with their own brilliance. The simplest of plants, these one-celled organisms have no roots, no stems, no leaves. Yet out of lifeless matter such as the carbon dioxide and nutrient salts they receive from their coral host, each can carry on the complex chemical process of making the sugars, proteins, and vitamins that nourish the animals of the sea.

Though this coral's life began when the sperm released into the water by one polyp reached the eggs formed in another, it needs no such mating in order to reproduce. Within itself it contains the seed of an entire community. Soon a companion polyp has budded on the tubular stalk and is building its own limy exoskeleton. And then a progression of buds open and mature. As the years pass thousands are grown, each joined to its

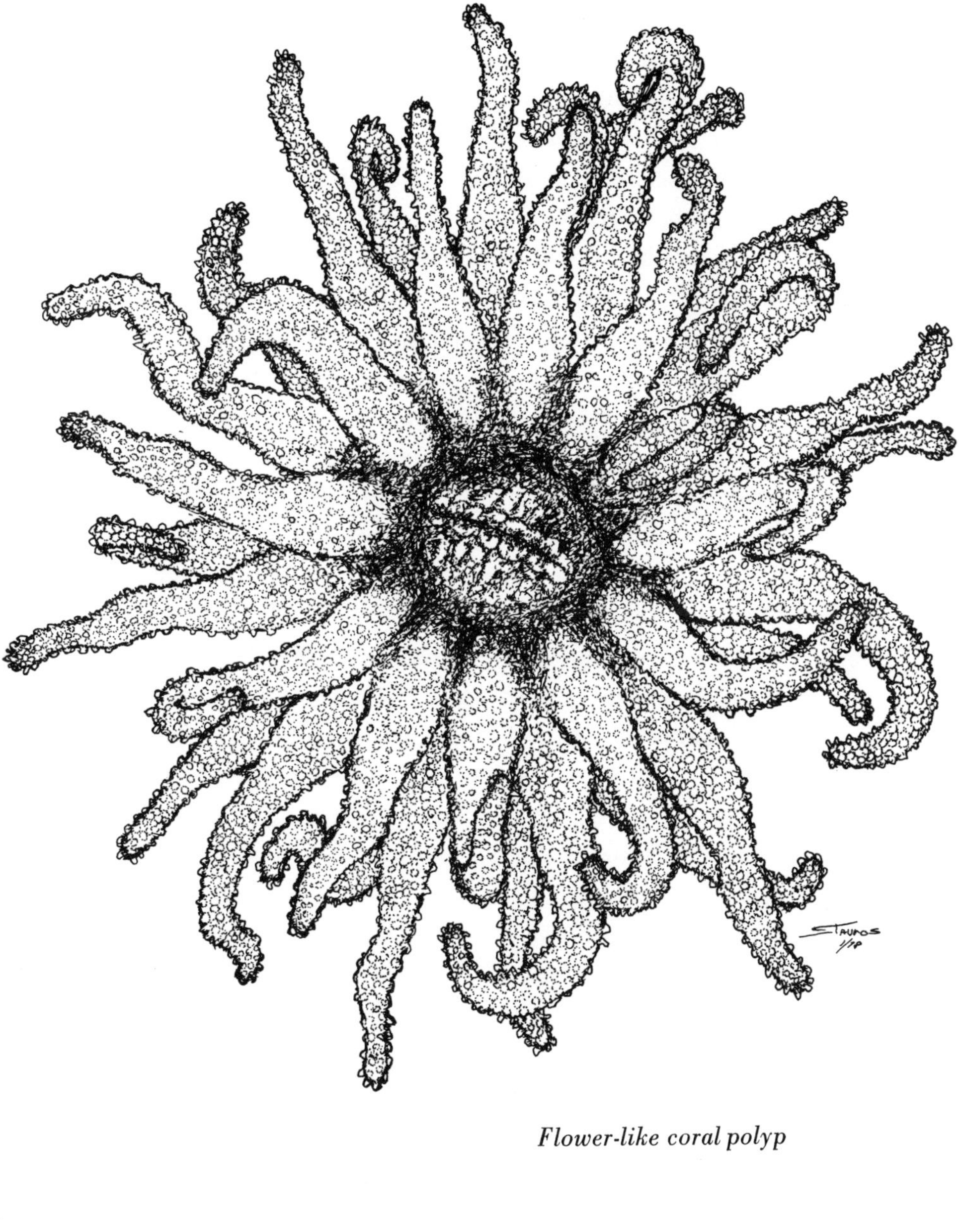

Flower-like coral polyp

parent and neighbor by a horizontal sheet of tissue folding outward just above the skeletal cup. From a single polyp no more than three millimeters in diameter a colony several feet in size has resulted, its individual members so intricately bound to each other that they even share a common digestive tract.

Under the best of circumstances, the coral polyp has a precarious existence. Like all sedentary animals, it is always in danger of being smothered by a sudden shower of silt. Dilute the salt content of its waters with a season of torrential rains and its development may be arrested. Let an unusual bloom of algae fill the surface layer of its sea and it may be poisoned. Lower the temperature of its shallows abruptly, and it is dead. But where the ship is resting, conditions are encouraging the budding polyps to thrive. Gradually the rusting decks and corridors of the ocean trader are being terraced by coral growths of intricate color and shape, the original colony having been joined on board by other larvae, each reproducing in a different design but always multiplying, crowding together, advancing upward upon the limestone remains of its dead relatives. Some circle irregularly into flat, spiny pads, while others mass into giant boulders, their coiling skeletal rows resembling the convolutions of a human brain. Still others thrust outward like the regal horns of a giant elk or branch gracefully like a bouquet of radiant flowers. Also flourishing in the coral shrubbery are the "soft" members of the family, the broad sea fans and slender sea whips, which are not protected by lime skeletons but cover themselves instead with a tough, leathery secretion.

In time, the wreck is further inhabited by an array of sponges and green algae that hangs from its twisted superstructure and decaying bridge like the moss that curtains a tropical forest. Snails and oysters and other shelled animals—mollusks of all descriptions—gnarl its decks and sides; shimmering anemones and sea squirts blossom beneath its hatches. In place of footsteps, its passageways are filled with the silent flutter of fish. Burgeoning with bumps and whorls, the vessel has already gained considerable height and the shape of its hull has been markedly changed. But centuries will pass before the corals overtaking it have grown to where their ramparts stand just below the break of the waves and its original outline is lost. For now, it is a glorious hulk protruding from the sand as though nature had mimicked human engineering by fashioning her newest reef in the guise of a ship.

Into this stony garden comes a young female octopus wanting refuge from the faint blue light dawning in the sea. Earlier she had discovered a barrel among the debris on the bottom—to her, a hole with high wooden sides and a powerful feeling of privacy, one strong enough to lure her away from the shell-built home she was returning to. But it was already occupied by another of her kind and she was driven off. Later this day, the octopus that chased her will be taken from the sea, and not even the rocking of his barrel as it begins to rise will induce him to leave it. Though startled by the unnatural lifting of his home, the creature will lack the impulse to escape. With no enemy visible, the idea of safety to which he has been bred will compel him to do no more than curl into a smaller shape and keep to the bottom of his hole. Held in place by unswerving habit, he will easily be brought to the surface and captured by fishermen who have learned to make a trap of his instinct for survival. But in the holds of the ship, the female has found an unmovable haven, and after her searching arms have assured her that nothing likely to challenge her is crouching inside, she follows them down into the soothing closeness of a more secret place.

For two months she secludes herself in the ship, each day without incident, emerging only at night and even then avoiding contact with any creatures other than her prey. But she has not come here entirely alone. Inside her body, near the base of the gills, she has borne packets of sperm fastened there by a male she crossed paths with some time ago. During her stay in the cellars of the ship these tiny caches have burst, saturating her with their seed, and now the eggs fertilized by them are just beginning to appear, each as minuscule as a grain of rice. Like the fruit of a vine, they come out in clusters on slender stalks, and this ivory spawn she hangs from the roof of her home. When she is done there are threads containing nearly 20,000 capsules draped around her.

With a single-mindedness that even causes her to

Octopus caring for her eggs

forego her food, she continually cradles these frail strands in her arms, gently brushing the capsules clean of any foreign substances settling on their surface. Occasionally she also washes them with gusts of water sprayed out of her body. All things capable of decay—including those she might eat—are methodically swept away. Other than this there is nothing she can do to aid the development of her offspring, for they are biologically self-sufficient in the form they have been spawned. But her one custodial task is essential to their survival. Those capsules deprived of a mother's hygienic fondling are most susceptible to invasion by parasites and fungi that can cripple and disease the babies and may even prevent their being born.

Twice as the babies grow toward emergence they reverse their position inside the capsule, the second rotation taking place just prior to their birth. Now the backside of each is pressed against the broader end of the membrane, ready to struggle out in advance of the head and arms when this cap finally lifts. All the while the infant is pushing at the hole widening in its prenatal sac, wild flashes of color are radiating through its body—a fireworks display proclaiming the intensity of its effort to be born.

Upon hatching, Onta, the baby octopus, and her brothers and sisters immediately abandon the ship for the open ocean. But instead of scattering about the bottom where they will dwell as adults, they stream upward, beckoned nearer the surface by the brightness of the water there and not the least perturbed at being surrounded by innumerable microscopic organisms that blow about in the currents like dust in a wind—the elemental plankton, swirling hub in the sea's wheel of life.

Wherever the water possesses the optimum combination of sunlight, mineral enrichment, and temperature, the various species of the phytoplankton flower. That they have never evolved from their unicellular form is testimony to the completeness of their primeval creation. A single cell, however small it may be overall, is larger in surface than it is in volume. This enables it to resist sinking and remain in those zones where the sun's rays are strongest. If the plankton cell had grown to greater mass, it would have had to develop a device such as the bladderlike floats some animals use in order to stay adrift. And still it would have lost advantage. Having the greatest possible surface area, the one-celled plant is also better equipped to absorb as much nitrogen, phosphorus, iron, manganese, and B vitamins as it needs for photosynthesis, especially when these substances are in short supply. Most of the diatoms and dinoflagellates of the phytoplankton are so small they flow with ease through even the finest screens. But what they lack in dimension, these wandering algae make up for in total population. Once their season has begun, they may triple in number in just a few days, tinting the oceans yellow and green for hundreds of square miles. These are the hues of a fertile sea, an aquatic plain where there is good grazing for the plant eaters and thus more food for the carnivores that follow them.

Throughout the winter, the ocean is layered with an abundance of fertilizing salts, but the plankton pastures remain thin, like burrowed seed on land, waiting to be primed by the spring sun. But unlike the meadow grass, they wane again in late summer. By this time the minerals in the upper zone have been exhausted and there is no upwelling from the bottom to replenish them. Under a flaming sun and becalmed skies, the surface waters have heated to the point where they are running in isolation from the cooler, denser pools below, like a separate sea. But before the phytoplankton goes dormant again, there is one more intensive bloom. Chilled by the graying autumn winds, the surface waters grow heavier and sink. As the winds gather and become gales, the sea stirs again, churning the great reservoir of minerals and salts deposited on the bottom. Upward into the path of the drifing plants come the flecks of earth, eroded rock, and volcanic ash washed off the coasts and scattered by the ranging currents, the decomposed remains of last season's life, and the excretions of passing animals. From the emulsion of air and water formed in the waves storming over the surface, the diatoms and dinoflagellates acquire the rich source of oxygen without which they cannot propagate. Thus catalyzed by the ocean's rhythmic turbulence, they will continue dividing until the sun is too weak to kindle them, though not always

Aftermath of a red tide

to the benefit of the animals they were meant to support.

Under certain conditions, triggered perhaps by an unusual shift in the prevailing surface winds or the sudden influx of excessive quantities of nutrients, a particular species of dinoflagellate will overwhelm an area with its reddish color—and a consuming poison. Just as the Bible describes the rivers of Egypt turning to blood in the time of Moses and destroying all life within, so does the sea become a stagnant pool of dead, decaying fish when a "red tide" strikes. First some of the plankton feeders fall victim to the toxic secretion of the dinoflagellates; then the predators which they in turn sustain are poisoned. Panicked by the sudden lack of uncontaminated fish, sea birds abandon their nests and young. Bottom animals such as eels and crabs struggle ashore, exposing themselves to an alien environment rather than remain in their now lethal waters. All along the nearby beaches, millions upon millions of the ocean's creatures lie rotting in the air. Flowing everywhere with the sanguine bloom of the plankton tide is the stench of death.

But in spite of their stupendous capacity for destruction, the planktonic poisons causing the various red tides may eventually be turned to human advantage. In recent years the scientists of a relatively young discipline—marine pharmacology—have been making progress toward refining an extremely poisonous dinoflagellate extract for possible use as a medicine. This substance, known as *saxitoxin,* is also responsible for paralytic shellfish poisoning in people who ingest it in fatal quantities by eating clams, oysters, and other mollusks which have been feeding on toxic dinoflagellates. Deadlier than strychnine, saxitoxin will kill a mouse in microgram doses though some mollusks are able to accumulate massive amounts of it, apparently without ill effect.

Depending upon how much of the poison enters the human body, its victim will begin to feel a prickly, burning sensation around the mouth within 5 to 30 minutes. This uncomfortable tingling quickly advances to the fingers and sometimes even the toes. Gradually, a numbness spreads from these regions into the arms, legs and neck. Though the person is losing his sense of touch and the ability to maintain other physical contacts with his surroundings, his mind remains clear and he can still feel pain. Growing increasingly nauseous, he experiences a sense of being weightless, as though he were sailing off the ground. At this stage, he is also losing control of his muscles and staggers about as if he were drunk, unable to speak. Second by second, it becomes more difficult for him to breathe; eventually, he will die of respiratory failure.

To the biomedical scientist, there is nothing unusual about refining a powerful poison into a therapeutic tool. Perhaps the best example of this is the present use of *curare*, the well-publicized plant poison from the jungles of Brazil, as a muscle relaxant during surgery. Since saxitoxin acts so specifically on nerve and muscle tissue, it can serve as a molecular probe, helping us add to our understanding of the processes involved in nerve conduction at the basic, cellular level.

Interest in the offshore world as a source of medicine traces back to the earliest civilizations. Five thousand years ago, the fish was regarded as a symbol of health by the Chinese, who had learned to use a variety of seaweeds in the treatment of throat diseases, particularly goiter. In ancient Greece, seaweed preparations were relied upon to calm everything from liver disorders to gout. And for a citizen of Imperial Rome suffering from a toothache, the remedy was a mixture of vinegar and the ground-up spine of a stingray. These primitive attempts at deriving pharmaceuticals from ocean plants and animals were essentially a hit-or-miss proposition, lacking progressive study and refinement. Certainly the iodine contained in seaweeds was helpful in treating goiter; but it is doubtful that the Roman following Pliny's dental instructions received much relief from his or her pain.

In practical terms, the science of extracting new cures for diseases from underwater life is still very much in its formative stages. Only a few of the drugs being prescribed by doctors today have been developed from aquatic sources. Yet the sea houses most of this planet's major varieties of animals, many of which possess virulent poisons. The superabundance of toxic marine organisms is an encouraging sign that many species may contain materials for killing pathogenic bacteria or restoring some elements of the body's physiological balance. With too many microbes already breeding strains which can resist the "miracle" drugs combed from the land's resources, medical researchers are being forced to probe deeper offshore for more effective antibiotics, anticancer compounds, and other therapeutic agents.

But perfecting a new drug from the sea is a particularly time-consuming and expensive process. Perhaps its easiest phase is the discovery of a biologically promising substance. This natural mixture must then be broken down into its component chemical compounds and its critical ingredient isolated. Now the potential drug must undergo extensive experimentation against a wide range of microbes, tumors, and viruses. Its effects on the nervous system, musculature, and major organs of test animals must be carefully measured and understood. At the same time, the exact chemical structure of the active compound has to be determined so that large quantities of it may later be synthesized. Then the proper dosage, one which will enable the drug to fulfill its intended function without producing harmful side effects, must be worked out. Finally, it must be tested on humans.

As much as several millions of dollars and a decade of experimentation may have to be invested before a new drug from the sea is ready to be marketed. For this reason, most pharmaceutical companies prefer to leave the basic research to the independent marine scientist. Only after very definite preliminary results have been obtained will they commit themselves to the enterprise. Unlike the drug firms, however, the biologist's primary aim is deciphering the complex nature of the life systems in the sea. Thus he or she is prone to concentrate on a particular organism because of some novel feature in its makeup, regardless of whether or not this might later prove to be of commercial pharmacological value.

Still, it is not uncommon for both interests to be served well by such studies. The discovery of an unusual fungus in the coastal waters of Sardinia, for example, led eventually to the development of a new family of antibiotics —the *cephalosporins*—which can conquer staphylococci capable of secreting an enzyme that destroys penicillin. Considering that less than 1 percent of all the marine organisms that may have the capacity to heal have been fully analyzed, it is more than likely that in the future we will free ourselves of illness primarily with medicines derived in some way from the sea. Perhaps the next one will be taken from the poison of a red tide.

One hundred feet below the spreading algae, Onta and the other baby octopuses are bumped and shaken by a stream of tiny animals that whirl and flash about each other like a downpour of shooting stars. This is the company they have been seeking—the constant stampede of the zooplankton. Of all their future homes, this one will be the most dangerous—adrift in a crowd of riotous appetites where life is so accelerated it ends for millions of creatures only minutes after it has begun. As it ends now for the unlucky copepod, not much older than Onta but considerably smaller, which has just collided with

Animals of the plankton

the octopus's mouth and as a result of this accident becomes her first meal.

Virtually every group of organisms inhabiting the sea produces young which are temporarily planktonic, kept aloft by powerful ciliated cells arranged along their microscopic bodies in a multitude of patterns. For the time it takes them to assume their adult form they are surrounded by creatures too numerous to count and of such bizarre creation as to seem unworldly even in comparison to insects. To escape being eaten by such fierce hunters as the shimmering arrowworms, the armored crustaceans, and the kitelike jellyfish that sail among them, these larvae and other developing animals must grow as swiftly as possible. But they are also compelled by nature to float with the plankton for the longest possible time so that their species may be widely distributed, which further insures their survival as adults. Those that do so with the greatest success are the ones most nearly transparent; the more invisible an infant is to the predators swarming around it, the better its chances of withstanding a slower rate of growth. Like the slivers and fragments of a shattered glass sky, animals devoid of color swim with the plankton herd in greater safety, almost as undetectable to each other as they are to us.

Though not completely transparent, Onta and her kin are guarded by the same kind of protective coloring worn by the herrings and mackerels that range near the surface to feed. Seen from above, their darkish backs blend with the blackness below, while their pale undersides make it difficult for a predator coming toward them from the bottom, with the brightness of the upper waters in its eyes, to single them out. In two or three months—the exact length of time depending upon the ocean temperature, which in turn affects her appetite—Onta will have grown too heavy to remain afloat in the currents. Until then, she is at the mercy of their flow, soaring in a whirlwind of motions that will either set her in the path of mouths wide enough to swallow her or continue steering her aside by the barest margins. Before she has settled on the bottom, most of her brothers and sisters will have perished, their fate being shared in other animal groupings by similarly incredible numbers of the drifting young. Of the thousand eggs laid in the plankton stream beside Onta by a single anchovy, no more than a few can be expected to survive.

From the depth they keep by day, most of the zooplankton rise each night through the water column, led by the herbivores that will crop the tiny plants awash on the surface. By the furious stroking of cilia or microscopic limbs they climb, expending enormous amounts of energy just to reach the higher grazing grounds. Protozoa, jellyfish, siphonophores, ctenophores, arrowworms, pteropods, and most of the crustaceans and young fish come up after sunset to feed, then drop down again at daybreak, some covering a vertical distance of 400 feet in a single round trip. What reason lies behind these daily migrations which cost so much in effort? For evolutionary selection to have resulted in so many different animals undertaking such an extreme exertion, it must be critical to their existence. To suppose otherwise is to accuse nature of being supremely wasteful—in effect, of working toward a world in which it is the unfit who best survive. An animal designed to spend the greater part of its strength on an activity of marginal consequence is a mistake soon forced into extinction.

Most scientists attempting to account for the vertical ranging of the zooplankton argue that these animals sink for the day to escape illumination and frequent the surface only when it is as dark as their deeper habitat and they are concealed from their enemies. But how then explain the numbers of them that are luminescent and glow in the waves like beacons advertising a feast for the baleen whales and fish that hunt them? It seems far more realistic to say, as Sir Alister Hardy does, that the migrations are nature's way of insuring that large numbers of the zooplankton are not trapped permanently in barren waters. The ocean currents generally travel faster at the surface than they do below; thus, at the height of one evening's climb, the zooplankton can be a mile or two away from that part of the upper sea they had sampled the night before. For animals capable of swimming only a few hundred feet in a twenty-four-hour period, it does appear far more advantageous for them to turn upward in order to radically change their surroundings than it does for them to shift from side to side.

However strenuous their rising is, it does enable them to routinely vary their feeding grounds. As the sea rotates above them, there is always the chance that yesterday's desert will be replaced by a prairie.

Of course, it would be less fatiguing for the zooplankton if conditions permitted them to cruise the lofty sea grass without having to descend. But the one-celled plants they seek have chemical defenses that make this impossible. Where the phytoplankton are dense, the combined effect of these substances within each plant is so pronounced that some grazing animals avoid the area and those that do enter stay but a short time. When the fishermen of the North Sea encounter large patches of *Phaeocystis pouchetii*, a brownish-green colonial algae that clings to their nets like clumps of glue, they know there will be no herring caught that day. In the Antarctic, however, this same algae is consumed by the euphausid krill, an animal of the zooplankton, which is in turn eaten by some penguins. Once passed into the bird's body by the krill, the acrylic acid present in this plant has a remarkable effect: It destroys bacteria, giving the penguin a gastrointestinal tract that is virtually sterile.

In nature there is no ultimate weapon. As plants developed ways of repelling their predators, some animals devised means of countering these new chemical barriers. A few were not only able to neutralize the substances being wielded against them, but also grew capable of using this material in their own defense. The slow-moving California sea hare, for example, gains added protection from its enemies by eating the red alga *Laurencia pacifica* and absorbing the seaweed's toxic compounds. The adaptive interplay between these two organisms has become so intricate that the sea hare cannot successfully reproduce unless the alga is near it. Those of its larvae that do not settle upon the red weed will fail to change into the shell-less sluglike animal.

To our disadvantage, we are most aware of the plankton drifts as a food source for the larger animals and tend to study them primarily in this context, ignoring the intricate ways in which the microscopic members of pasture and herd fence among themselves for survival. Part of the problem is entirely physical. Some of these organisms are simply too small to be captured, even by the most sensitive devices we can employ. In one instance, though, planktologists tracking the tiniest species received unexpected assistance from an animal just 1/6 of an inch in length—the *Oikopleura*—which has developed an ingenious method of trapping creatures even smaller than itself. By means of a thin, transparent secretion, this tunicate seals itself in a domelike structure within which it also weaves two conical nets that empty into its mouth. With the flitters of its tail, *Oikopleura* then creates a current inside the dome that draws in fresh streams of the most diminutive flagellates and protozoa, which are filtered from the flow by the fine mesh of its collecting nets. While examining an *Oikopleura*, one scientist picked through the particles adhering to its nets and found among them flagellates measuring just 1- to 2/1000 of a millimeter—vessels of life so tiny no human-made screen could have held them.

Many of the sea's smaller inhabitants have never been kept alive inside the laboratory—another handicap in the effort to understand them. Before 1967, for example, the tintinnids that abound in Lower New York Bay had never been nurtured for the revealing eye of the electron microscope. Because no one had ever examined them in depth, they were classified solely on the basis of the size and shape of their loricas, the shell-like covering in which each of these unicellular ciliated organisms is encased. But when Dr. Kenneth Gold of the New York Aquarium's Osborn Laboratories of Marine Sciences discovered how to sustain these animals, it became possible to establish their place in the food chain and ascertain the circumstances under which they prosper. Now he wants to know how the creature forms its lorica. To accomplish this, radioactive isotopes are fed to the tintinnid and then traced to see if they are deposited in its shell. Eventually, Dr. Gold will expose this organism and others like it to various pollutants in order to gauge the effects on their growth and reproduction. If the laboratory tintinnids show a specific response to a particular substance, their kin in the ocean can be watched for signs of that pollutant's presence in critical proportions, enabling us to sound the alarm before its accumulation is so great that larger animals are stricken.

An appreciation of such minute organisms as the tintinnids is essential to any attempt at conserving the more imposing creatures of the sea. Unfortunately, the plankton do not generally receive the attention they deserve. Because they are so small, they escape our notice and too often are treated as though they are insignificant. When animals as majestic as the great whales are hunted to the brink of extinction, public concern for their preservation can be aroused; but mention an avoidable threat to a plankton species and only a few scattered voices are heard repeating the cry for its protection. Yet collectively these specklike creations are irreplaceable in the food chains that lead to the giant mammals. To work with them requires the patience of a master craftsman, the ability to scale one's perceptions down to the infinitesimal without also shrinking the sense of wonder, the elation we experience when witnessing the existence of creatures closer to our own size. But until we are sure of how the plankton relate to each other in their own private sphere—in a chain of life that at its height is still nearly invisible—we risk not being able to preserve them for the benefit of those animals more impressive to us because they are more easily seen.

Varying in size from minute unicellular forms of a few microns in diameter to the large seaweeds that are many meters in length, marine algae are not only the primary and major producers of organic matter in the sea but also offer substances that are used extensively in the food, cosmetic, and pharmaceutical industries. Kainic acid, obtained from a red seaweed, has given doctors another remedy for intestinal worms. Alginic acid, also a product of algae, is the stabilizing ingredient that is added to liquid medicines so that it is no longer necessary to "Shake Well Before Using." Calcium alginate, a seaweed salt, is woven into surgical dressings because it builds a gel that arrests bleeding. This same quality makes alginate wool an important aid in dentistry. Alginates can also be used for both preventing and treating radioactive strontium poisoning. When toxic strontium enters the body via food particles, it can be combined inside the gastrointestinal tract with the alginate to form an insoluble gel, which then passes out harmlessly through the bowels. If the radioactive strontium is already deposited in the bones, it will be resecreted into the intestine where it is bound by the alginate and fed into the feces. Other seaweed extracts are helpful in cancer therapy. Patients with bone metastases having an algal phycocolloid administered to them after surgery or x-ray treatment show improved appetites, are free of nausea and fatigue, and have less painful symptoms than are normally associated with this particular cancer. Certain acidic polysaccharides found in seaweeds are known to reduce cholesterol levels by preventing the body from absorbing this substance after it has entered the intestines. Some of the seaweeds commonly eaten in Japan have also demonstrated an ability to lower the amounts of cholesterol present in the bloodstream of laboratory animals. If these benefits can be extended to human beings, then the sea will be providing us with materials that may help protect us from the follies of diet and life-style by which we now make ourselves increasingly susceptible to heart disease.

Fishermen from the harbors of the North Atlantic, both in America and in Europe, can always earn a bit extra in summer by rowing their dories out to where the seaweeds grow and harvesting whole boatloads of the type known as Irish moss (*Chondrus crispus*). This reddish plant, which flourishes on rocks or other hard surfaces in shallow waters, contains a substance—carrageenin—that is a key ingredient in the processing of thousands of everyday products, especially foods. A highly effective food preservative and stabilizing agent, this versatile extract holds liquid mixtures in solution, thickens creamy solids, giving them a smooth and uniform body, and helps block oxidation, thus retarding spoilage. These properties make carrageenin a valuable additive in the processing of a host of dairy products, beverages, bakery goods, sauces, meats, and fish. It is also a vital component in the making of molds from which dentures, as well as other forms, are cast. There is even a use for it in insect sprays and deodorants. In the United States alone, the production of carrageenin is valued today at more than $10 million annually, making it a substantial industry; yet the seaweed that supplies it is still harvested throughout the world in much the same way as aquatic plants were gathered for food, medicine, and fertilizer some 5,000 years ago.

Looking like an odd-colored clump of parsley, Irish

Gathering Irish moss

moss commonly grows in height from 2 to 12 inches and attached to rocks just below the low-water level, though it is also found at depths as great as 30 feet. From the stalks that rise out of the holdfast (the small disc at the plant's base), a number of thick branches emerge at their tips, giving this seaweed a crisp and tufted appearance. Its precise color depends upon several factors: the time of year, the chemical composition of the water, and the amount of light reaching it. Through summer, the Irish moss growing in shallow water with good exposure to sunlight comes up yellowish green; in autumn, this same moss turns brown. Those plants fixed in deeper water or in a greater degree of shade turn an iridescent reddish purple. Where there are huge beds of it, this seaweed covers the tidal rocks and ledges like heather on a hillside; closely packed and sturdy, it stands as straight as a soil-grown plant when the sea recedes and it is open to the air.

In some regions of Canada, people wait for a storm to wrench the Irish moss loose from its holding and beach great heaps of it on the shore; mostly, however, it is brought in by people who rake its sturdy branches from the rocks or pluck them off by hand. It is a task that requires skill and must be carefully done. The pull of the rake, which usually has an 18-foot handle, is all that keeps the dory in position. If the plant's holdfast is damaged or dislodged by harvesting, there will be no further crop there for several seasons. There is also the danger that other vegetation may prosper or that mussels may become attached to the space previously occupied by the holdfast. Considering the small size of this plant and the crudeness of the tool with which it is grabbed, it is easy to see how an inexperienced harvester can ruin a moss bed. But a person who understands the nature of the plant and its proper gathering can safely rake in up to 1,000 pounds during a single tide.

From May until September, the raking continues along the shores of New England and Canada, each region having its own time for harvesting, depending upon the weather. Generally, long stretches of sunshine in the spring will bring on an early harvesting, while cool and cloudy days will delay it.

Once the moss is brought ashore it must be dried, unless it is being sold wet; otherwise it will spoil. This is best done by spreading the branches out in a thin layer on a flat, clean area such as an unused paved surface. Beaches and meadows are less suitable because they make the moss dirty and it dries more slowly on grass and sand. In some places, a stretcherlike platform made of boards nailed together between two side rails and known as a "flake" is used for the drying. With a good wind and not much moisture in the air, the moss should dry in one or two days. During this time, the red pigment in the leaves becomes so concentrated and dark that the plant is called "black moss" by the men who handle it. If the seaweed is to be bleached as well, it must be resoaked and kept moist with salt water for as many as three days, the branches being turned regularly to allow them maximum exposure to the sun. But during rainfalls, the branches must be covered to prevent their being washed by the fresh water, which will leach out their natural salt content. Extracting the carrageenin is done by first rewetting the Irish moss, then spinning, filtering, and purifying it with activated carbon. Any water remaining is removed by evaporators and the solid extract is recovered.

Though an important industry in other parts of the world, Irish moss gathering is not done extensively in the United States, which recently has been importing about seven times the amount of seaweed harvested from its own waters. But this may change in the near future as a result of a major breakthrough in marine farming techniques being made at the New York Ocean Science Laboratory in Montauk, New York. In a project that began seven years ago, three of the laboratory's scientists, Dr. Henry W. Moeller, James P. Hunt, and Steven Giordano, have managed to seed Irish moss in their tanks and raise the plants to maturity. And the moss has grown much faster under controlled conditions than it does in the sea, where plants of this type take as long as three years to develop. This is partially because the seaweed grows in its natural environment for only a few months each year, but in a laboratory setting it will continue sprouting throughout the year. Most important, however, Dr. Moeller and his associates have learned during the course of the experiment which genetic strains do best in their system.

Dr. Moeller's first attempt at farming a seaweed other than Irish moss was made in 1967, using a 200-gallon cylinder filled with air which was anchored to the seabed about 5 feet below the surface; arms radiating up from the cylinder like the spokes of a wheel were covered with a netting on which the seaweed was planted. For three years he experimented with this method and achieved a surprising result: although the seaweed did well, mussels also lodged on the netting and they were thriving too. In fact, these mollusks eventually did better than the seaweed. The model was so successful with regard to the mussels that it was later refined into a method for cultivating them that is now being used commercially. Fertilizing the seaweed had been Dr. Moeller's greatest problem. Since the fertilizer diffused throughout the open water, it was too expensive to use in the quantities needed to insure that the proper amount landed on the plants. In addition, too much of the natural debris floating in the water was settling on the seaweed and shadowing out its light.

In 1970, he moved his work onshore and began raising Irish moss in his own tanks. He specifically chose this seaweed for his new experiments because it already had commercial importance and there would be immediate practical use for whatever lessons were learned. At first he grew the plants in the kind of plastic pools that parents buy for their children to splash around in. Though the moss did well here, keeping it alive during the peak months of the summer was difficult. This was overcome by shifting the experiment into a series of concrete tanks through which cooling water could be pumped when the temperature outside went too high. In this system, the tanks are arranged in three rows of five each, with the second row lower than the first and the third lower than the second. Seawater for the tanks is pumped out of a bay adjacent to the laboratory and delivered to a holding tank. The water flows down from the holding tank through the growing tanks and ultimately back into the bay. Another problem partially solved in the new tanks was that of the epiphytes, the small plants that grow on other marine vegetation. Dr. Moeller found that by keeping the density of the Irish moss at the highest possible level, it dominated the environment to the extent that it progressed at the expense of any plant life around it and the epiphytes' growth was retarded.

Having demonstrated that Irish moss can be farmed with laudable success, Dr. Moeller is now concentrating on improving the stock, endeavoring to farm one that will combine the fastest possible growth rate with the greatest amount of carrageenin. He believes that because of the complex molecular structure of this substance and the ease with which the seaweed containing it can be raised, it is far more practical at this time to concentrate on farming Irish moss than to have chemists attempt to synthesize carrageenin in their laboratories. With projected improvements in his present method, he estimates that Irish moss can be grown to yield an annual crop worth some $40,000 per acre. Moreover, this seaweed also possesses a sulfated polysaccharide called *ebimar*, which is a proven and effective antacid, giving it added commercial value.

As blood pulses through the human body delivering to the cells in its most extreme quarters the chemical ingredients vital to their functioning, so do the aquatic currents course about this planet mixing animal and plant communities and the raw materials that sustain them, thus insuring that the distant oceans do not become stagnant pools of isolated life forms. Driven by the wind and differences in water density, these subsea streams are also greatly affected by the earth's rotation, whose motion they are compelled to match in speed. With the ocean's surface water flowing at all times in an angular direction to the prevailing winds, a considerable pressure may be exerted on the fluid layers below. This downward thrust bellows the deeper water out at the sides, slowing its rotation by increasing the area it encompasses. No longer spinning in harmony with the seabed, the bulging band of water must either speed up or veer toward the equator, where the earth's rotation is progressively slower. If it moves on, as it most often does, its place will be filled by a segment of the sea which is following a polar heading as a result of its gyration having been accelerated when it was squeezed upward to replenish the waters at the surface being steered away by the wind. It is this constant surging of the oceans, tuned to the spinning earth, that produces the great currents like the Gulf Stream and makes it possible for seeds from tropical plants native to South and Central American shores to be found on the beaches of the Outer Hebrides, a family of balding islands off Scotland's west coast chilled by mists as thick as curling smoke and as damp as fine rain. Not surprisingly, those ocean animals patterned to spawn at points where the dominant current leads their young into relatively safe nurseries are the ones that are most likely to endure.

Had the infant octopus Onta been born close to the tropical region of the Atlantic just northeast of the West Indies where the sargassum weed gathers, she would now be treading next to a brood of flat creatures that look like sheer silken leaves shaken out of an invisible tree by the passing aquatic breeze. For two and a half years these tiny fry will drift near the surface before they reach their journey's end in the river mouths and brackish coastal waters of Europe, North Africa, and the Middle East. By then in their third summer of life and having grown to a length of about 80 millimeters, they will lose the tiny sharp teeth and transparent appearance that once caused them to be mistakenly classed as a species of planktonic fish and assume the shape of young freshwater eels—ready to leave the sea and swim upriver. Though *Anguilla anguilla*—the European eel—may live as long as eight years, it mates but once; and for its moment of reproduction, many scientists believe it must return 3,000 miles to the place of its birth. Waiting in the coils of its genetic code as it matures is the signal for it to transform from a yellow-toned inhabitant of inland waterways to a deep-ocean voyager, silver in color and possessing the enlarged eyes characteristic of creatures that cruise the lower depths.

A similar migration to the Sargasso Sea is also embarked upon by the American eel (*Anguilla rostrata*), which would be indistinguishable from its European counterpart if it did not have fewer vertebrae. Like the sargassum weed itself, which is ripped by storms from the coastal rocks of Central America and the West Indies and collects in the enormous gyre that shares its name, the eels departing from American rivers mass here to seed another of their brood. It takes six months for the weeds torn from West Indian shores to float up to the northern border of the Sargasso Sea, several years more for them to be carried into the inner rings of this eddy. But once in the calm of the Sargasso's center, they are virtually immortal and will continue to grow, though without rocks to attach to they cannot reproduce. Bobbing above the courting eels, the sargassum fronds are a vivid yellow, unmarred by epiphytes because of the antibacterial substances the weed produces when healthy.

Hatchlings of the American eel dispersed in the currents are launched so that none strike the mainland below the equator. But south of their territory is another creature referred to as an eel (though it is actually a fish, kin to the carp and goldfish)—a misnamed prowler of torpid creeks and streams whose capacity for generating an electric charge powerful enough to immobilize a person at close range made it a prime subject of biomedical research in the early years of World War II. At that time, it was still feared that nerve poisons might be used in battle and an effective antidote for these doomsday weapons was anxiously being sought.

An animal's nerves will quit functioning if they are deprived of the enzyme *acetylcholinesterase.* Knowing this, scientists had developed poisons that rendered the enzyme inactive by saturating it with organophosphates. What remained to be discovered was a quick and safe method of releasing this indispensable substance from its lethal phosphate bond. For the scientists at the New York Aquarium and Columbia University responsible for delivering this solution, the electric eel was an invaluable source of research material. Throughout its snakelike body there are thousands of specialized cells circuited to emit two types of pulses: the first, a harmless 10-volt discharge which the animal uses to scout the waters ahead in much the same way as a whale relies upon its sonar; the other, an eruption of 500 volts that might stave off an enemy or knock out its prey. A system producing such intense electrical activity literally flows with acetylcholinesterase. Thus, from a relatively small number of these creatures, the research teams could extract as much of the enzyme as they needed to complete their work. As it happened, the nerve poisons were never released by either side during the war. But the antidote developed to counter them has since been used to revive people accidentally dosed with insecticides that also kill by means of organophosphate poisoning.

From another creature of the sea, a marine worm commonly used by Japanese fishermen as bait, we have derived an insecticide that eliminates pests without using the organophosphates so harmful to humans and other animals. Having seen that some carnivorous insects died as soon as they touched this annelid, Dr. S. Nitta isolated its active ingredient in 1934, a substance he named *nereistoxin.* While it blocks ganglionic action in an insect's central nervous system, this poison does so without inhibiting cholinesterase, making it nontoxic to warm-blooded animals. After 33 years, during which time more than 100 chemical derivatives of this substance were screened, it was ready for marketing under the trade name of Padan. Farmers have since been able to use it against insects that are resistant to organophosphates and organochlorides without fear of harming other plants and animals in the environment. Padan also has the advantage of decomposing fairly rapidly so that it does not add to the mounting chemical contamination of the world.

Of all the migrations undertaken by creatures such as the European and American eels who must cross huge expanses of water to reach a special breeding ground, few end in a visible display of mating as spectacular as that of the pelagic squid. Propelled by the force of water jetting from its funnel, the squid rockets through the sea like an errant missile, its sleek, luminescent body flashing first one way then another, yet never seeming to be slowed by these sudden changes of direction. Unlike its close relative, the octopus, whose habit it is to crawl the ocean bottom in cautious solitude as an adult, this cephalopod is a creature of the open sea, sometimes swimming alone but most often ranging the major ocean currents in large, rapacious schools.

Its body long and tubular like that of a fish and tapering at the rear into either an end or side fins, the adult squid varies in overall length from specimens as tiny as a few inches to monsters of sixty and seventy feet. Eight slithering arms it has for grappling with its prey and, in addition to these limbs, two rubbery tentacles that wave about in the water like the antennae of a giant insect. At the very end of their reach, these whiplike appendages bulge into the shape of clubs, which are studded with the same kind of suckers that cover the entire length of the creature's arms. The clutching disks of an octopus grow flat against its limbs and have a smooth surface, but the squid's sit atop a short, muscular stalk and are rimmed with tiny needles.

In most of its species, the squid speaks to the human mind with an exhilarating appearance of color and speed; and with its arms gathered in and pointing ahead of the rounded eyes like a prickly snout, it does indeed have the gift of lightning in its underwater flights. For some even the surface of the sea is no barrier to the course they fly. Like a volley of arrows aimed upward by an army of phantom bowmen, whole schools of squid will shoot for the roof of their world at such tremendous velocity that they soar through it into the open air. As they do, their body fins and the membranes on the outer edges of their arms spread as wings to stabilize their aerial ascent, and they arc over the waves until the momentum of that last underwater thrust from their funnels is spent and return, still in perfect formation, to the sea.

To an approaching danger the squid can react with split-second quickness, accelerating almost instantly to high speed. Its ability to do so is explained by the unusually large size of some of the nerve fibers linking its brain with the muscles responsible for this motion. In animals, as a general rule, a nerve impulse travels fast or slow depending upon the diameter of the fiber conveying it—the messages carried by the thicker ones taking less time to reach their destinations in the body than those being transmitted through thinner relays. To the squid's great advantage, evolution has shaped it so that the further a nerve fiber must extend from its brain to connect with some portion of the mantle, the thicker the fiber becomes. Thus, all the muscles in the mantle—even those at its extremities—are being flashed by the creature's brain at virtually the same moment and can therefore respond to whatever command has been given with the full thrust of their combined force. Since it is the mantle that powers the animal's movement by driving water through the funnel as it contracts, the simultaneous operation of all its muscles makes the squid a thing of such exciting speed and agility that even its most ordinary maneuvering is a fantastic sight.

The existence in this animal of nerve fibers measuring up to one-twentieth of an inch across has proven extremely beneficial to scientists striving for a more precise understanding of how nerves function. First identified in 1909 by Leonard W. Williams, the squid's largest nerve fibers, or axons, are some 50 times thicker than those possessed by humans and most other mammals. With material of this dimension in hand, physiologists have been able to do things, such as planting electrodes in the living fibers, that had previously been beyond the bounds of their technology. And because the biochemistry of nerve conduction is essentially the same in all forms of animal life, the lessons grasped in studies of the squid's axons could also be applied to humans.

For some reason, however, investigations involving the giant nerve fibers of the squid did not begin in the years immediately following their discovery—perhaps because, as Williams himself inferred, the scientists of that day simply were not ready to believe that anything so large could really be an axon. But in the 1930s, these exceptional fibers were rediscovered for neurophysiologists by J. Z. Young. His more detailed descriptions of them placed particular emphasis on their importance as test materials for future research, and now the experimenters being given this information were prepared to make use of it. Through their work, which others continue today, the squid serves humankind as a mirror in

49104

which we see more of the nature of our own nervous system with each probing look.

Not many animals hunt the elusive octopus once it has begun its life on the bottom, but the squid has a legion of enemies—predators as different in description and habit from each other as they are from the prey they commonly seek. Pulsating jellyfish, Antarctic penguins, seals, sharks and other large fishes, and the sperm whale all share a taste for this creature. From all except the mountainous whale, the giant squid is protected by its size and strength and ferocity. But the smaller species must utilize much subtler weapons to counter their numerous foes, relying exclusively for their defense on the mosaic camouflage of their pigment cells, or chromatophores, the momentary concealments provided by their inky discharges, and the incredible outpouring of speed that makes them seem capable of vanishing when attacked.

Yet in the turbulent jungles of the ocean, not even a creature blessed with phenomenal swiftness and chameleonlike disguises is assured survival. Those of the smaller squids that manage, under ordinary circumstances, to elude their natural enemies and evade the fisherman's net must face still another danger—a compulsion from within that cannot be avoided and that for a time each year makes them deadly to themselves and to all others of their kind.

By the hundreds of thousands, then the millions, they

THE LIBRARY
GRAYS HARBOR COLLEGE
ABERDEEN, WASH. 98520

swarm in coastal waters during the spring and summer months, urged into these great gatherings by their obsessive need to mate. Since they first began to close on the spawning grounds, not one of them has hunted and fed itself. Now that they are here and wedded without exception to the pursuit of one another, they are oblivious to whatever else is happening around them. Sharks and whales and other predators whose arrival nearby would normally send the squids flying for their lives swim in their midst and swallow glutinous mouthfuls of them at will, yet provoke no alarm. Twisting about in a frenzy of seeking and grasping, the squids have no sense of anything but themselves and do not flee even though they are being eaten in great numbers. Swimmers of magnificent grace and symmetry, they streak the sea now in wild disorder, their moment for embrace a scrambling chaos that erupts upon the surface and churns it to a froth.

Until they are joined, the squids of both sexes are pale and shimmering, the color of cobwebs; but as the male's arms play over his shackled mate, a darker pulse beats into each limb. Sporadic at first, these crimson flashes build with rhythmic intensity to a brilliance the arms will hold for as long as the couple remains entwined. One arm of the male is specially adapted for passing spermatophores from the organ of his sex into the female. Should her eggs not be ready for laying when she is pursued, she will face her mate directly so that his encapsulated sperm can be attached to a place on her buccal membrane. Afterward, the individual sperm cells

Squid mating

will seep from their pouches and collect in a cuplike depression not far away on the membrane. There they will remain inactive until the laying begins. A female whose eggs are ripe, however, will be seized in a different way. While the couple is swimming—she with obvious nervous excitement, her arms cast apart and wobbling as she spurts forward—the male follows closely at her side until all the signals of their courtship have been displayed and the two are ready to touch. Then he swoops beneath her. From this position he will reach for her body, anchor it in his grasp, and enter her through the mantle cavity with his special limb. For about ten seconds this arm stays inside, ejaculating spermatophores whose masses of fertilizing cells are covered with a sticky substance that holds them in place beside the opening of her oviduct. When the sperm from these packets is released, much of it is lost, washed out of her through the funnel like a dissipating, ragged cloud. But the number deposited within is so large that despite this loss enough cells remain to quicken her eggs.

Their contact at its natural climax, the squids of this pair are nevertheless unsated. During the next few hours they will mate again several times, though not necessarily with each other. In the fever of their coupling, the male may also try to feed his other hunger by eating parts of her body. Some males are so aroused they insist on thrusting into the nearest swimming form even though this squid is one of their own sex and resists violently. Still other males are pairing off to fight over a female both have been pursuing.

For many of the females in this thrashing, diaphanous tide, however, the desire for union is spent and the last of their energy is being concentrated on securing a proper place for their spawn. Encased in masses of jelly, the eggs are laid, each gelatinous cluster containing about a hundred of them. A single female may produce as many as twenty of these soft, cylindrical cases. As each appears, she cradles it in her arms for a few minutes, then dives straight to the ocean bottom. Feeling about for an object to which she can attach the case, she bounds forward on the tips of her arms as though they were metal springs. When finally she locates a suitable rock or clump of algae, she presses the string of eggs against it. The subsequent hardening of the collection's outer jelly glues it to the chosen surface. This is all the female squid will do for her progeny, but it is enough. So distasteful is the protective surface of the case that few animals will bother with it.

In just two or three days, billions of encapsulated eggs have been rooted to every available outcropping in the region and to each other, clustered in bunches like an immense array of blossoming flowers, the great mass of them covering more than 200 acres of the sea floor. A month or so more and they will hatch a new generation to swim these waters in place of the one that is dying now from the unbearable strain of having passed on life. Those females still able to move are nothing like the full-bodied, energetic creatures that flooded this sea a short time ago. Their mantles torn and flaccid, they weigh half as much as they did when the mating began. The males that ravaged them are also thin and suffering, too weak to go on swimming. At the spectacle's end, the individuals of both sexes falling like a silver rain over the garden of life waiting on the ocean bed to be born.

II. Afternoon

Blue whale harpooned

From the moment of her birth, Onta the octopus has closely resembled her parents in appearance. For her the attainment of adulthood will not be celebrated by a sudden and radical transformation in body function and design, the seemingly magical change from infant larva to wholly different adult experienced by many ocean creatures, such as the barnacle which turns in distinct physical stages from a free-swimming, translucent creature to one encased in a ridged shell shaped like a lampshade and fastens itself to another object for the rest of its life. Outwardly, Onta's progress toward maturity is evidenced by a simple, straightforward growth—the gradual altering of dimension rather than an instantaneous recasting of shape. It is in demeanor, the inner aspect of emotion, not morphology, that the developing octopus undergoes a metamorphosis. Drawn at first to the sunlit regions above, Onta becomes more and more discomforted here as she grows larger. Once content to drift in a cyclonic swarm of plants and animals, she is beginning to feel burdened by too much company—the nuisance of things that get in her way and do not fully satisfy her expanding appetite. With her size and weight steadily increasing, she cannot rely on the currents to keep her aloft in the plankton cloud; it is usual now for her to expend considerable amounts of energy just to maintain position. This adds to her vexation. After long periods of swimming she must glide and rest, but this is impossible without some sinking, which ultimately prods her into action again. Each day the routine of straining to remain in the glaring brightness, the relentless crowd becomes more oppressive; in time it will be so distressing that Onta will feel compelled to seek relief below. On her short excursions out of the plankton stream she has begun to know the solace of being alone, an experience so tranquilizing that the desire to repeat it is taking command of her urges. Haunted by too much exposure, she senses an escape from all her torments in the darker waters pointing down toward the bottom.

Like a celestial visitor orbiting an uncharted planet, Onta cruises the middle waters looking for the best place to touch down. Far from being a series of smooth, unchanging plains, the ocean basins are areas of intense ferment bearing every geological mark and feature found on land—deep canyons, giant mountains, ridges and plateaus, erupting volcanoes, and vast stretches of prairie rivaling, and in many instances surpassing, their landed counterparts in height and breadth. Eager as she may be for the bottom life, Onta can stall her descent while she is over unfavorable ground—just as the larva of other animals ready to end their planktonic phase can delay their physical metamorphosis until the world below is home. Unobstructed sand flats or high, vacant peaks, for example, do not tempt her into settling down. Though one of the conditions driving her from the aquatic sky is her reluctance to share space with any creatures—even those that cannot harm her—Onta is in pursuit of a particular kind of isolation. For the sake of her appetite she must exist within another community, so she hunts now for the rough bottom terrain that will provide her with a suitable cave in which she can closet herself until it is time to eat. Before she will commit herself to a landing she will have to sight the tumbled foothills of an undersea cliff or the cellar haunts of a coral palisade or some other sturdy concealment inside a realm inhabited by hordes of other animals. And there she will dwell like a fugitive, the tenant whose presence is occasionally heard and sometimes felt but who is seldom seen, stealing out only to prey on her neighbors.

Passing above the octopus as she scans the lower region are some of her relatives who will never cease to drift—the pteropods, snails whose bodies flare out at the sides in winglike extensions. Resembling butterflies that do not dip forward or backward but flutter just fast enough to remain in flight, these tiny mollusks will go on spending their energy struggling to stay in the stream until they are either picked off individually by other microscopic challengers or cropped in great bunches by a giant plankton-eater, like the 35-foot-long basking shark which is coming up behind them now. Its familiar fin breaking the surface and attracting a squadron of seabirds expecting to be led to a meal, the shark is maintaining a steady pace of four kilometers an hour as it steers for the plankton stream. Instinctively, the hungry animal holds this speed though nearing its food, for going faster would have the effect of forcing a column of water ahead of it which would push aside many of the tiny organisms it would rather be swallowing. Like the exhaust of a mammoth engine, water taken in through the shark's huge mouth pours out from its gills while it feeds. In a single hour more than 2,000 tons of water will be filtered by the shark, the plankton sifted from it being caught in the gills by a special raking mechanism shaped like a comb.

More than a few of the sea's larger creatures are linked for survival to the smallest animals in the food chain. Among them are the toothless whales, the *Mysticeti*, who are able to support their stupendous bulk by feasting directly on the zooplankton; and the grandest of them all is the blue whale, larger and more powerful than any other animal that has ever lived on land or in the oceans. Frequently measuring more than 100 feet in length, this colossal cetacean may weigh as much as 170 tons. Its tongue alone is heavier than a full-grown elephant. A massive, heavily muscled tail capable of generating up to 500 horsepower with its motion gives the blue whale incredible strength. One female, after being harpooned by the crew of a modern 90-foot catcher boat, is said to have towed her attackers 50 miles in an eight-hour ride, though their vessel's twin engines were pulling in reverse all the way.

After migrating from its mating grounds in more temperate seas, the blue whale spends the summer gorging itself on the tiny, shrimplike krill that abounds in the cool waters fringing the Antarctic ice. Barely three inches long at its biggest, the krill is so richly concentrated in the south polar sea during this time of strong sunshine (when its own principal food source, a greenish diatom, also flourishes) that its swarms color the water reddish-brown for hundreds of square miles. And just as well too, for the blues returning to harvest this crop are painfully thin, having had virtually no food for the eight months they were away. Cruising slowly near the surface where the krill cluster most thickly, a famished whale takes in a tremendous mouthful of brine. Raising its tongue, it forces the water back out through the overlapping plates of baleen, or whalebone, which hang instead of teeth from its upper jaw like a great strainer. Another flick of the whale's tongue sweeps the horde of minute organisms trapped in the baleen curtain down its throat. Then its immense mouth swings open again and the feast continues.

Once the whaling captains realized that the blues always came back to the Antarctic in November and December in anticipation of the summer spree (which begins in earnest with the arrival of the New Year), it was a simple matter for them to converge on the area at the appropriate time and ambush the preoccupied whales. A blue surprised by a ship while grazing might not even notice its assailants until after it is wounded; suddenly immersed in a banquet, it is so absorbed in the wonder of being able to eat again that all sense of danger is lost.

As intelligent as these cetaceans are, it was impossible for them to alter their habits after the whalers began decimating them. A pregnant female must leave the Antarctic to have her calf, for the coldness of the water there in winter would kill the baby. Though the newborn blue is about 25 feet in length and weighs almost 3 tons, it lacks the protective layers of blubber needed for insulation against the subfreezing temperatures of a polar winter. When the expectant female departs, she is accompanied by her mate. The blue whale has no feeling for being part of a herd, but a mating pair will stay together for life, traveling from year to year as a solitary family. While they journey in less chilling climates, their calf is growing at a fantastic rate, adding an inch and a half and 200 pounds every day on nothing more than its mother's milk. Seven months after its birth, when the infant is fully weaned, it measures 53 feet and is some 20 tons heavier. During this time, its parents have been kept alive by the accumulated fat of last year's summer-long meal. In every sea they have crossed, the plankton concentrations have not been dense enough to consistently fill their stomachs and enable them to maintain their proper weight. Where else but the Antarctic and the waiting catcher boats could this family have gone to keep from wasting away? Why should they have defied nature and chosen certain death by starvation when they were the lucky ones who had always been out of range or had otherwise evaded the whaler's lance in the past?

In 1965 there were so few blue whales left alive that only twenty were brought in from the Antarctic. That same year the International Whaling Commission finally outlawed all hunting of this endangered creature in its polar feeding grounds; but the commission may have acted too late to save the blue whale from extinction. No one knows how many of these animals have been overtaken outside the Antarctic, or slaughtered illegally there since the ban. Given the attitude of most whaling nations toward species preservation, it would be miraculous if there has not been significant poaching. And even if the killing had ceased as ordered, the blue might still be doomed. It is estimated that less than 200 remain, and these few survivors are scattered from the Antarctic ice to the edge of the tropical seas. Some must be juveniles, not yet ready for reproduction—others too old to bear new life. Those capable of fostering another generation of calves must first locate each other, and the chances that they will are questionable at best since they search in a territory covering more than one-third of the planet's surface.

Too few parents and a low rate of reproduction for those that do mate are not the only obstacles to recovery for the blue whale. This gargantuan cetacean also has an aquatic enemy capable of eliminating it entirely now that its numbers are so reduced. Considering the enormous potential for destruction contained in a beast the size of the blue whale, it is indeed difficult to imagine it becoming the victim of any predator other than man. Nevertheless, there is one native of the sea that dares to stalk this leviathan; a creature so savage and cunning in its assaults on ocean mammals that it has been named "Killer."

As an unsuspecting blue serenely swallows acres of the sea's plankton harvest, the menacing black and white figures of a pod of *Orcinus orca*, their sharklike dorsal fins riding high above the water, are rapidly closing in on it. Distant cousins of the giant they are about to attack, the orcas, or killer whales, are among the smaller cetaceans, reaching a length of no more than 30 feet and a weight of only 7 or 8 tons. But instead of baleen for filtering zooplankton from seawater, these whales have mouths lined with sharp, conical teeth for tearing flesh—up to twenty-eight in each jaw, hooking inward and interlocking like the crippling prongs of a steel trap. Also unlike the blue, which swims alone when unmated, the orca travels and hunts in packs, as many as fifty individuals cooperating at times on a kill. Though it is more likely to attack penguins, seals, baby walruses, and dolphins than a giant cetacean, the killer can be tempted into challenging a blue whale if its chosen opponent seems weak and its desire for a delicacy is especially strong.

Suddenly the blue is aware of the threat bearing down upon it; but although orcas are appearing all around the frightened whale, it has become virtually motionless. Even after two of the orcas have aimed themselves directly at its head, this Goliath is too intimidated to stir itself and avoid their charge. Clamping their powerful teeth into the blue's lips, the two killers begin wrestling its jaws apart. Still the beleaguered whale makes no attempt to break free, fight back, or at least churn its immense strength into a bulldozing run for safety at its best speed of 20 knots.

While the lead orcas rip at the mouth of their impotent prey, their companions slash forward repeatedly to bludgeon the tiring giant with their tails and tear at its sides. Without letup the onslaught continues until, finally, the blue is too weak to keep its jaws clenched any longer. The inside of their victim's mouth exposed at last, the killers have their prize: the whale's great tongue. Gorging themselves on it and then swimming off, they leave behind a tortured blue that may still be alive—a helpless derelict slowly being consumed by the scavengers attracted to the slaughter.

So terrified are some baleen whales at the mere presence of the killers that, when surprised in a bay or inlet by a pod of orcas, they have literally committed suicide by hurling themselves on shore, preferring to die from prolonged exposure to the sun and air rather than be eaten alive. Only on rare occasions will a larger whale use its size and strength to ward off the killers. In one such case, a humpback whale met five orcas head on in defense of his mate and calf. Lashing out furiously in all directions with his punishing tail, the enraged humpback drove his would-be destroyers in search of easier prey. But however often this may happen, it will not outweigh those instances in which the giant is defeated. And one day the whale being hounded to death may be the last blue ever seen, finished off by its native enemy but only because it was pushed to the edge of oblivion by human hunters.

Who can say exactly how many species nature herself has eliminated through climatic change and geological upheaval, and the perpetual striving for genetic advantage that has existed since the first organic molecules competed in the primordial ooze for food and living space? Certainly far more than human civilizations have exterminated—more even than the total number of species existing today. Whatever form life takes, it is predestined to contend for survival with both the chemistry of its surroundings and the capabilities of its nearest rivals—those organisms constituted to thrive at its expense. And it is axiomatic that neither the geophysical forces of the planet nor the patterns of contesting consciousness can remain static; nothing of our perceiving is allowed to remain as it once was, though we may believe it so because it changes at a rate too slow for us to comprehend. Somewhere in the world the high, jagged peaks of a comparatively young range of mountains are being smoothed by the constant rub of wind and snow, their pointed heights interminably chipped and buffed and washed away so that in time they will be as round as hills and almost as easily climbed. But what a time it will take for that to happen. In a thousand years, human eyes will not even notice a difference in the sharpness of the crags, as they do not see (nor do we feel) the continents drifting apart, the Atlantic basin growing larger as the Pacific shrinks.

If an organism is critically late in adapting to a developing challenge or does so inadequately, it disappears, and there is no reason why this should be regarded as inherently wrong. Would the triumph of these stricken plants and animals through a different shuffling of circumstances necessarily be right if that wished-for restructuring of earth, sea, or sky (or their populations) dealt a death blow to some other species? In the sea, for example, wherever a cold current passes beside a warmer flow one may find planktonic animals that are so sensitive to temperature they will border each other but will never cross out of their respective waters and mix even though they may be traveling no more than a boat-length apart. Remove the thermal wall that separates these species (as the physiochemical processes of the planet have in the past, and are likely to do again) and one group will be overcome while the other prospers. Whether the shift is to a generally warmer or colder sea, at this level the changing temperature is as much a killer as it is a benefactor. How then should we judge which way the barrier should fall?

Illustrating the difficulties of applying the principles of conservation to a world of diverse interests should not be taken to imply approval of the concept that human beings are as much an instrument of natural selection as a critical shift in the magnetic field or the explosive birth of a volcanic island and therefore cannot be held accountable for what they do—that whatever the act, it is the natural consequence of human domination of this planet and produces a new reality that is neither better nor worse than the one it replaces, merely different. Ascribing this character to our behavior defines us in essence as being no more than exceedingly clever animals, never bound to think beyond the satisfaction of our own senses. Yet one of the distinguishing features of higher consciousness is an individual's willingness to voluntarily widen the sphere of his or her heartfelt concern, extending it from oneself to one's mate and close relatives and even further into the realm of strangers, reaching as far as our nonhuman neighbors. Once this truce is established between a person and the totality of his or her surroundings, that individual's idea of what constitutes a personal gain or loss is likewise expanded and a more thoughtful and harmonious management of the environment becomes possible.

No matter how cautious we are, though, we cannot proceed on this planet without in some way interfering with the course of other life forms. In many areas and aspects, the landscape we are anxious to preserve because we regard it as being unspoiled is by no means free of human influence—wild in the way it was before prehistoric tribesmen began torching sections of woodland rather than settle and farm wherever an indiscriminate bolt of lightning had cleared the trees. What we deem "natural" is usually relative to how we see the land just before someone proposes altering it again. But while we cannot avoid having an effect on the appearance and composition of the biosphere, we can carefully gauge the changes being sought, resisting those that would be undertaken without due regard for the consequences to other organisms and their living space.

In every situation a proper balance can be drawn between society's need to use and yet somehow preserve the world's physical endowments. Finding the compromise fair to both points of view has never been as difficult as persuading people of the fundamental importance of taking the time to look for it. The future may prove that most human beings are emotionally unsuited to making a realistic assessment of what an unrestrained level of consumption will cost the world and unilaterally tempering their appetities on its behalf. But we are all capable of realizing that our most selfish interests are often best served by lending vital support to someone who will then be obliged to repay this debt at our bidding, or by protecting an animal or plant population that contributes in some measure to our being provided with valuable raw materials. We do not have to be conspicuously self-sacrificing in order to help another person or a separate species survive. We simply have to understand and accept that our own well-being is ultimately tied to the welfare of those creatures who cannot exist for very long if we do not make room for them in our plans.

Had the policymakers of the major whaling nations been moved by such enlightened selfishness, the blue whale could be hunted today without question of its extinction. Studies made by the International Whaling Commission (which nevertheless failed to protect the animal in time) estimated that a population of 60,000 blues could have yielded 6,000 adults for the factory ships each year without declining. Instead of acting as responsible businessmen concerned for the future prosperity of their enterprise, the whalers have been ravaging their stock as though it were inexhaustible, ignoring all reports that it was not. At its peak, the blue whale population was probably never higher than 150,000; yet in a single year (1931) 30,000 of these creatures were taken. Other baleen species have suffered the same treatment as the blue, succumbing to men who defiled the art of commerce by massacring whales as others like them once murdered buffalo, leaving an unnatural void in the pattern of life because they had devastating power and were wholly lacking in vision.

If one considers the world to be a single integrated organism containing within itself tinier pockets of life arranged by species in a multitude of cooperative subsystems, then humankind too often deserves to be likened to the cell that goes berserk and begins overwhelming all those around it. One by one the ecosystems which fuel the planet's ebb and flow are decimated by this rebellious cell's unrestrained lust for growth. Some it poisons with its alarming amounts of metabolic wastes, others it swarms across and engulfs, but all seem destined to be sacrified to its blind march forward. Like a cancer in the body, we are capable of contaminating so many branches of the earth with our capacity for unnecessary destruction that this vibrant, intricate organism we inhabit might sicken and die. Fortunately, we have not yet advanced so far that the elemental forces of nature can no longer contain us. Conceivably the weather could worsen, as some predict it will in the years immediately ahead, exposing us to extremes that will blight our food supplies, increase the spread of disease, and otherwise batter our delusion that we were designed to reign supreme. Still, we can be extremely stubborn and slow to learn. Once we have recovered from the calamity of being thrown back into harsher times, the shock of being struck with physical proof of how vulnerable, perhaps temporary, we are and have accustomed ourselves to the new shape of our seasons, we might again be dreaming about conquering the world.

It is this capacity for questioning the harmonies of the earth and being audacious enough to believe we can retune them to strains of our own imagining that awards us special distinction among the animals, not merely the ability of each human to perceive that *I am.* A growing number of behaviorists now feels that we are not the only form of consciousness that can acknowledge individuality by separating *me* from *them.* It seems likely that however modified or reduced an animal's awareness of itself may be, it has some inner sense of standing beside, if not actually aside from, its companions. Chimpanzees have shown that they can not only learn to use human sign language, but can also recombine the original word images they are given to meaningfully describe novel occurrences—in essence, to think for themselves and perhaps pass the information on, which may signal the beginning of their climb into sophisticated communication. But we can do more than merely recognize the difference of things. We alone seem to possess the spirit to ask *what if* . . . and the craft to subsequently hammer and bend the realities of this world until they conform more closely to our image of how they ought to be. Other species do not tend to create as many problems for their environments. If their desires or habits are misleading, they hurry toward extinction but do not compel those who are innocent of their error to follow behind. We, on the other hand, do affect the fate of every creature in the biospheres we insist on recasting according to our own purpose.

Our deepening intrusion into the underwater province has endangered more than just the blue whale and some of its cetacean cousins. Ten years ago we apparently failed once again to appreciate in advance the fine and fragile balance nature draws between opposing species and may have helped unsettle the way in which a spiny sea star and the fabled reefs of the Pacific coexist. In a single night one of these animals, commonly known as the crown-of-thorns, can devour a coral colony that has taken half a century to grow. With its population on the march, a herd of these coral-eaters can destroy an entire island reef in a meal lasting but a few years; in less than three, the species devastated many miles of the coral fringing Guam. Normally, the crown-of-thorns forages in the dark, crawling over the coral slopes on as many as sixteen arms; but when congregated in thick bands it sometimes fed on the polyps during the day as well, its digestive juices dissolving them inside their limestone cups, where they hide from the moist heat of the sun. Wherever the rampaging sea stars passed, the rainbow coral became bone white, the color of its death.

From Borneo to Tahiti, a succession of island reefs and atolls were stormed by the sea star's advancing hordes. If the reefs had been totally destroyed, the islanders would have been deprived of their main source of food. Gone in search of other sheltered precincts would have been the broad and bountiful pyramid of marine organisms intimately associated with a coral sea; and free then to bear down upon the beach, the ocean would have begun carrying away the shoreline of each unguarded island. Even a coral formation the size of Australia's Great Barrier Reef was in danger of being badly mauled by the invading crown-of-thorns. Between Cooktown and Townsville, large sections of the reef were turned to ruins, a graveyard of whitened stumps gathering a crust of algae that would prevent any new coral embryos drifting down from taking hold.

Usually, coral colonies help check the population of the crown-of-thorns by eating great quantities of the sea star's larva while they float about the reef. But if a sizable portion of the coral suddenly dies, then the larva in that area will mature at a much higher rate. By consuming the coral living on the borders of this sanctuary, these unexpected adults will then widen the space in which the next generation of larva can safely develop. Until the supply of polyps has been exhausted, the voracious sea stars continue increasing; but by the time they have migrated or are starving, the reef could be past repair.

Crown-of-thorns devouring coral polyps

What was our part in this tragedy? Some naturalists have suggested that we have been carelessly murdering large segments of coral by fishing in lagoons with dynamite and using machines and explosives to smash new and bigger passages through the reefs. At several of the besieged islands the crown-of-thorns was first noticed on the rise in places where blasting and dredging were going on. Our tampering with the coral seascape also includes depleting it of the triton, a giant snail which is the natural enemy of the adult crown-of-thorns. The spiral shell of this mollusk is prized as a decoration, and it is believed that as many as 100,000 were lifted from the Great Barrier Reef alone in just one decade. Though shell collecting may not have precipitated the problem on the Australian reef, it most certainly was a factor encouraging the crown-of-thorns to multiply. Another theory lays the blame for the initial death of the polyps on the insecticides and other pollutants which are building to troublesome levels in the open sea as well as in coastal areas. The haphazard deforestation of some island might also have been a contributing cause. Without enough trees to hold the soil, too much earth is washed into the ocean by the seasonal rains and the surrounding coral is buried under an avalanche of silt.

Uncertain about the primary reason for the crisis, biologists could only grope for its cure. Raising tritons on special farms and then releasing them wherever they were needed was one alternative that at least promised to restore healthier numbers of this creature to the sea. At some islands, the sea stars were met by teams of scuba divers armed with hypodermic guns for injecting lethal doses of formaldehyde into their ranks. One by one, tens of thousands of sea stars were systematically poisoned. Eventually, ecological equilibirum seemed to be brought back to the infested areas. But we cannot be sure that it will be maintained in the future since we are still meddling with the biological mechanisms of the reef without first having cataloged and understood the complex associations that define the life of a coral community. It is possible that what we have considered a "plague" of sea stars, a monumental threat to the coral by a species being given unfair advantage, is actually a normal occurrence in the life cycle of a reef—something we neglected to notice until the mid-1960s because we were just then learning to pay serious attention to the condition of the oceans. The most recent analyses of the situation have indicated that the crown-of-thorns generally crops the faster-growing corals, which are not so easily overgrazed, and by their selective feeding help clear space for those corals slower to mature, thus enabling the reef to maintain diversity. Also, not all of the reefs that were attacked were picked clean and rendered incapable of regeneration. In this sense, the sea stars are not unlike a small forest fire which opens the way for a greater variety of species to reestablish themselves in a revitalized soil. Seen in the light of these suppositions, the specter of thousands of spiny animals lumbering forward in a tightly banded group becomes less menacing, and we must wonder what damage we are doing to the natural order of things by killing them en masse as though they were an enemy of nature. As yet we do not know enough about the crown-of-thorns to explain how it migrates or even what its normal numbers are in any one location. How then can we presume to say that their breeding has been out of control? What makes it unnatural for a particular reef to wane because of animal competition while others are left alone or are capable of recovery? Nothing is meant to endure forever in one place and in one immutable form. And surely this intended sparring is not the same as our destruction of the coral through the use of chlorine bleach to flush whole droves of fish from their hiding places in the reef. Though outlawed, the bleach is still being employed, the polyps this poison cloud touches dying not from honest mistake but as victims of mindless greed.

To expect all material and technological progress to halt until every particle of knowledge has been gleaned from an environment is to deny the realities of human existence, our own legitimate will to press on. But allowing business to proceed according to dictates that show little or no regard for the ecological consequences is equally foolhardy. What we may ultimately have to rely upon as our guide for determining whether we have sufficient information about a particular life system to risk entering it for our own benefit is the legal construct of the "reasonable person"—the one who takes an appropriate amount of care to see exactly where he or she is going and what the step-by-step effects are of getting there. Just how much concern and foresight that involves must naturally vary according to the details of each situation. What is critical, however, is that both scientist and industrialist have in mind at the onset of their endeavor the question of how much caution goes into being reasonable, and then be honest with themselves about how well they are honoring this restriction. And if error is unavoidable we should become accustomed to committing it on the side of the animals, for each time we race our technological aspirations too far ahead of our biological knowledge we chance upsetting a wonder of life that may be truly irreplaceable.

Before we mark ourselves too indifferent to preserving organisms that do us no harm, we would do well to reevaluate our concept of intelligence, for it is in the smugness of our assumed superiority that the heart of our problem lies. We have always prided ourselves on being the masters of thought. Our idea of the crowning mental achievement is the one we can best do. All other species are compared against this self-determined standard, and because we are examining their capacities in our terms, they are invariably found lacking. Nowhere is this practice more misleading than in the world offshore. Living underwater makes the offshore creature as alien as if it were in fact a being from another planet. To be sure, this strange zone has its share of sluggards; but then, it is only logical that we also expect a few of its creatures to possess a commanding form of genius. Trying to discern it requires that we put aside all flattering thoughts of our own evolution and wonder how beings immersed in a liquid sky, who have neither the means nor the need to conquer more than their stomachs can hold in a single feeding, would be likely to express their brightness. We must accustom our eyes to looking for a mental excellence that has nothing to do with the construction of machines and the tearing down of a natural setting.

It was more than fifty million years ago that the ancestors of today's whales left the land, giving these mammals more time to develop in relation to their environment than we have had. For most of that time, the cetacean brain has been comparable to our own in size; but we have been at this peak for only 100,000 years. A killer whale has more neocortex than a human, and this is the part of a mammal's brain associated with its ability to react to sensory information above the basic reflex level. Those areas of the brain in which are located our power to reason and decide, to devise and govern social relationships, evaluate our own behavior and resolve to improve it—the highest centers—are also very large in the cetacean brain. It may also be fair to presume that in the whales they are at least equally as well developed. We do know that most whales are extremely social animals, communicating with each other by means of a complex system of sounds resembling squeaks, clicks, and shrill screams. The ability of dolphins to discriminate with their echolocation mechanism between different objects of identical thickness and sonic reflection empowers these animals with a cerebral talent we can only match through the use of computers and other highly refined electronic devices. Since most cetaceans have few enemies and their food supply is both ample and readily obtained, it may be that they have extensive periods of leisure that their acrobatic gamboling cannot fully satisfy. In such reflective hours, how might their minds be applied? Dr. John C. Lilly of the Human/Dolphin Foundation has seen that individual dolphins will, of their own accord, struggle to duplicate human speech. If this is not merely imitative but indicative of a desire to converse with us, mammal to mammal, what might we expect them to tell us of their race and its impressions of the world? Do they have an oral history, a legend of creation, expressions of poetry—knowledge that we might find humbling as well as enlightening? Dr. Roger Payne of the New York Zoological Society has discovered that the humpback whale actually is a singer of songs, repeating sounds ordered in phrases and themes bearing musical consistency, each individual performing its own distinct composition. Is this an operatic way of speaking or just an entertainment or perhaps something of both? We do not as yet know. One thing is certain, though: The cetaceans we maim and destroy so shamelessly are a breed much more ancient than our own, and however far they have progressed with their magnificent brains, they at least have had the good sense to do so without warring against the creatures beside them or attacking the sea itself.

Gliding nearer the pitted vaults of a barrier reef, Onta has escaped the baleen giants sweeping the plankton from the sea above like vacuums sucking in dust. She is also fortunate that she will never grow large enough to be individually hunted by the sperm whale being driven this day to the outer depths by a special craving that a bigger octopus might have satisfied. Deeper than the light of the sun can penetrate, the great bull dives in its hunger. For some time it has been feeding in depths where the water is tinted green or purple by the day, taking as much as 300 pounds of fish in a single meal; but now its urge is for a different, more elusive prey. Down beyond the scattering light it pursues this succulent thing that swims in dungeon blackness near the ocean floor—down, further down, until it is 3,000 feet below the air it breathes. More than an hour will pass before the whale has fought and killed its rare prize, yet in this time it will not return to the surface for another breath. From the last filling of its lungs it must derive enough oxygen not only to maintain consciousness for as long as it is under, but also to drive its body through the furious battle waiting at the bottom of its dive. In all, it is a herculean exertion that no mere expansion of the animal's lungs to full capacity could accomplish. Something is happening inside the whale, some marvel of its physiology that enables a mammal of the sea to hoard the oxygen in its blood and keep from weakening even toward the end of a marathon hunt.

Barely a century ago marine biologists discovered the vital nature of this compensatory mechanism, termed by them the diving-reflex phenomenon. Since then, however, their understanding of it has done much more than merely explain how whales, seals, and other aquatic mammals can stay submerged for periods of time that seem to us almost interminable. The investigation of this process is also proving beneficial to those studying the onset and treatment of shock in humans, as well as instances of sudden death for which no physical cause can be found and which had, in the past, automatically been attributed to heart failure.

The instant the sperm whale dived, the circulation of its blood was restricted exclusively to its brain, heart, and lungs, these being the organs that cannot do without an adequate supply of oxygen for any appreciable length of time. If the animal's brain were deprived of oxygenated blood for just five minutes it would suffer severe and irreversible damage. But other parts of the whale's body, such as its muscles, can endure this loss and continue functioning for hours, so they are cut off from the flow at once, as the blood vessels feeding into them undergo immediate constriction. Since its circulatory system has become drastically smaller, the whale's heart must now be slowed down so that the pressure within its active veins and arteries does not mount to dangerous levels. This too happens the moment the animal closes its blowhole and begins to plummet; the frequency of its heartbeat decreasing until now, as the whale searches the dark ocean plain with the clicking and pinging noises that are its eyes underwater its heart is pumping blood at *one-tenth* its normal rate.

In this state the whale is a potent hunter, a life more finely tuned to its environment. But when similar changes take place in the body of an ordinary human, there are normally no advantages gained; in fact, if the condition is not quickly corrected, that person will most likely die from the debilitation known as shock. A body falling into this state experiences an immediate slowing of the blood. This in turn has the effect of lowering the amount of oxygen being delivered to the cells until, finally, there is no longer enough for their metabolic needs. Unless it is reversed, this anoxia of the tissues will progress to a point from which there can be no recovery. Naturally, the body has already been acting in defense of itself, trying to maintain an adequate blood pressure and conserve oxygen for the brain and heart by closing off the vessels that normally pipe blood to the limbs and skin and such organs as the liver, kidneys, and intestines. But if unduly prolonged, the constriction of peripheral blood vessels cease being an aid to the body and instead accentuate the anoxic state of the tissues, making it that much more difficult for the person to come out of shock. The longer any part of the body is forced to go without oxygen, the greater the build-up in it of lactic acid, carbon dioxide, and other harmful metabolites. Ultimately, a point is reached where these substances are present in such large amounts that they cannot be handled. There are also indications that the hypercoagulability of the blood in the early phases of shock may subsequently hinder the victim's recovery since it heightens the possibility that vessel-blocking clots will be formed.

In humans, the vascular constriction experienced during shock often persists and becomes irreversible *even after* the factors that initiated it have been treated and brought under control; yet a similar condition in the whale and other marine mammals is thrown off by these animals with routine ease each time they surface for air. Why this is so is not presently known. What little we do understand about the neural physiology and blood chemistry of these animals suggests that at least one element of the explanation may be a special modification of their enzyme systems which facilitates the elimination of harmful metabolites. It also appears that aquatic mammals have a substance in their blood that protects them from intravascular clotting while they are diving but does not interfere with those mechanisms that promote the rapid clotting needed when the animal is wounded and bleeding externally.

The studies made to date in these areas are but first thoughts on the subject and raise more questions than they answer; nevertheless, their importance cannot be overstated. Once we can say for certain how the whale and its companion creatures are able to resist impairment while undergoing severe vascular constriction, we will also have acquired promising new ideas for the alleviation of shock in human beings.

A few investigators are also attempting to draw a parallel between the diving reflex and certain causes of death in humans that are otherwise unexplainable. Thus, it has been theorized that in cases where an individual seems to have died as a result of extreme fear or the mere threat of being cast out by his or her group—the essence of the voodoo ritual—that person's death may have been helped along by an exaggerated reaction of his or her body identical in many respects to the diving reflex. That humans also possess the capability for a nonharmful diving reflex is illustrated by the masters of yoga who, by consciously lowering their heartbeat and centering their blood circulation on just the vital organs, can survive for hours though sealed in coffins that have been submerged in water. One doctor has offered clinical evidence that activation of the diving reflex in humans trapped under waters colder than 70° F. allows them to be revived with no mental damage even though they have been in the process of drowning for as long as half an hour or more and are brought up cold and blue with no apparent pulse or heartbeat.

From the echoes of its sonar ranging, the sperm whale now learns that its quest is nearly over. Swimming just ahead in this veiled, delusive sea is the thing it so dearly wants: a pulsating mass of boneless flesh measuring about 20 feet from the tapered end of its tubular body

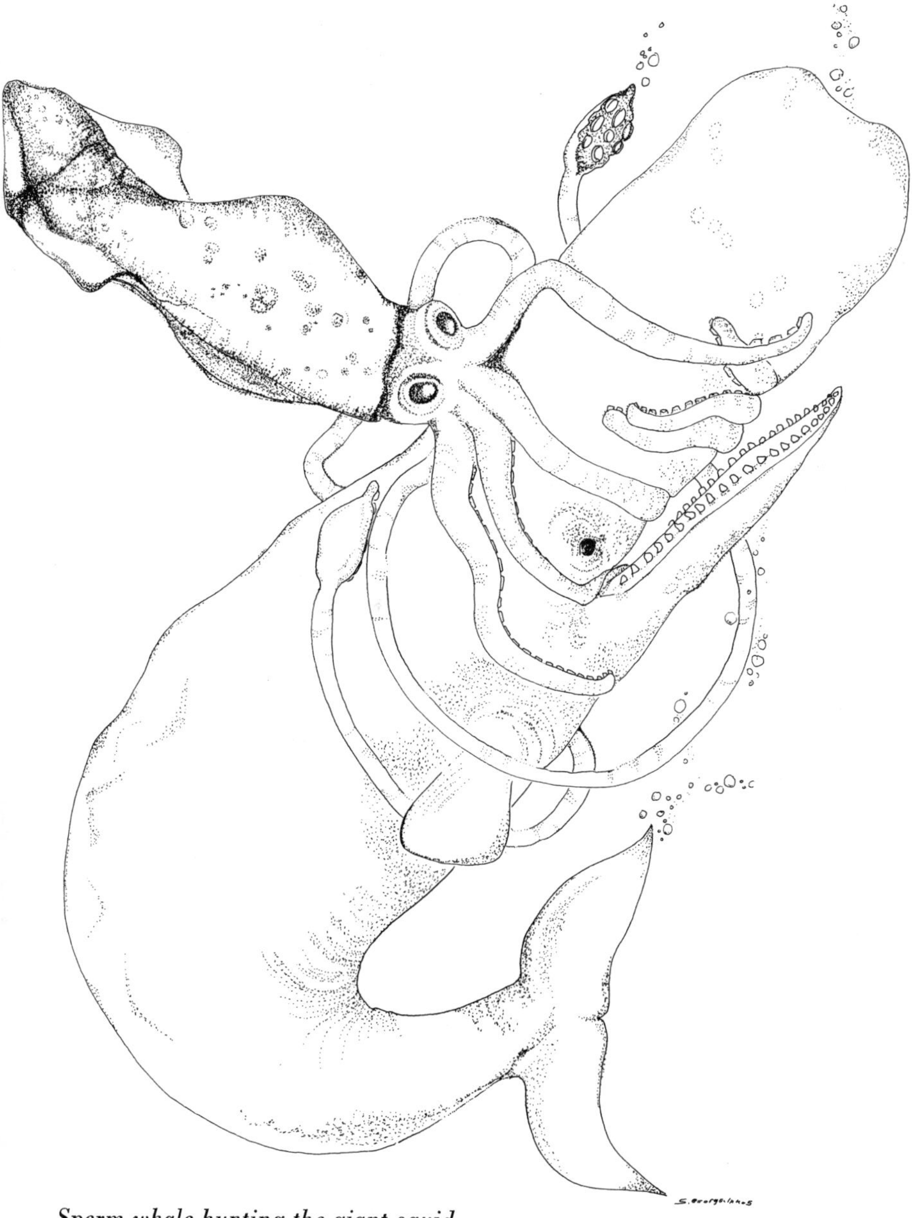

Sperm whale hunting the giant squid

to the sharp mandibles of its horny beak, with eight arms circling its head like a thatch of vipers, each limb stretching forward an additional 20 feet and as thick around at its base as the thigh of a stout man, and two slender tentacles that wave about like the antennae of a gargantuan insect and extend 15 feet further still. To the few humans who have seen it on the surface staring blackly at them with eyes almost 2 feet in diameter it is an unspeakable horror, this stupendous squid. But the whale is also a giant—60 tons of blubber and muscle that make the mere thrashing of its tail a weapon of enormous power. Inside the long and narrow sluice of its lower jaw are sixty teeth—huge things several inches long and weighing as much as 2 pounds each—that will rip through strong bone as well as flesh. It is a creature used to dominating, and even now it does not hesitate to attack.

Sensing the danger, the squid fires its ink, intending the looming discharge to attract and hold the bull's attention for the second or two it needs to escape—but the deception does not work. Still following the sound images of its sonar, the whale veers with the startled beast, raking the fleeing body with its spiky teeth. A moment more and the crushing jaws are clamping around the mantle of the squid, and once their biting grasp is secure the whale breaks for the surface. Swiftly it rises through one pressure zone after another without fear of the nitrogen in its blood dissolving into gaseous bubbles and causing a fatal embolism, as would surely happen in a human who did not give his or her body time to adjust to each change in pressure. But the squid will not allow itself to be taken without a final fight. As the whale climbs for air, the wounded mollusk winds its limbs around its captor's head. It has no hope of strangling the whale; but with the stinging suction of the muscular cups on its arms and the club-shaped tips of its tentacles it may be able to inflict enough pain, especially in the area of the blowhole, to force its release. All the way up the whale's skin is tortured by the stabbing pull of thousands of needle-rimmed suckers, some as large as 3 inches across. And now the squid's beak has emerged from its pulpy hiding place inside the soft head to snap and tear at the enemy's flesh—but to no avail. As the grappling giants breech, the jaws of the whale close down on the squid's brain, finishing its life.

A tremendous geyser of used air, water vapor, and mucus caps the whale's victory as it breathes for the first time in just over an hour. Now it must repay the oxygen debt of its body. All the tissues that were by-passed by its curtailed circulatory system and thus forced to retain the waste materials they empty into the blood as a matter of periodic routine must be paid off for their special labors with an abundance of oxygen. For a while the tired bull attends only to its breathing, rapidly sucking in rich blasts of air and expelling them in streaming spouts. Because it changes more than half of the air in its lungs with each breath (a human replenishes no more than 20 percent), the whale is soon at ease, once again able to dive deep and long. Only then does it begin tearing apart the great carcass floating at its side.

Arranged in about 150 species, the octopus has adapted to the peculiarities of nearly every sea, whether its waters be temperate, tropical, or polar. For the young adult reaching bottom in the Mediterranean, one of the strange creatures it must coexist with is a female sea worm that resembles a plant stem springing from a tailed bulb and branching at its height into two hornlike tips. During reproduction, the larvae hatched into the water by this echiuroid are undifferentiated and can become either males or females. Which sex they finally attain depends upon whether or not they come in contact with her long, tubular proboscis. Those that do will be exposed to an unusual growth- and sex-regulating pheromone secreted by her proboscis and body wall, and within a month they will have evolved into dwarfish males. Their growth permanently stunted, they will then live in a quasi-parasitic relationship with the female to which they are attached. On the other hand, the larvae that fall to the sea floor will grow into females and reach a length of nearly one foot.

Interest in the inhibitory effects of this critical secretion, known as *bonellinin*, prompted the biologists of the Osborn Laboratories to test it on sea urchin eggs and on cancers. Because this disease involves an unrestricted multiplication of cells, the discovery of a new compound active in reproduction always gives rise to the hope that

Bonellia viridis, *source of a possible anti-cancer drug*

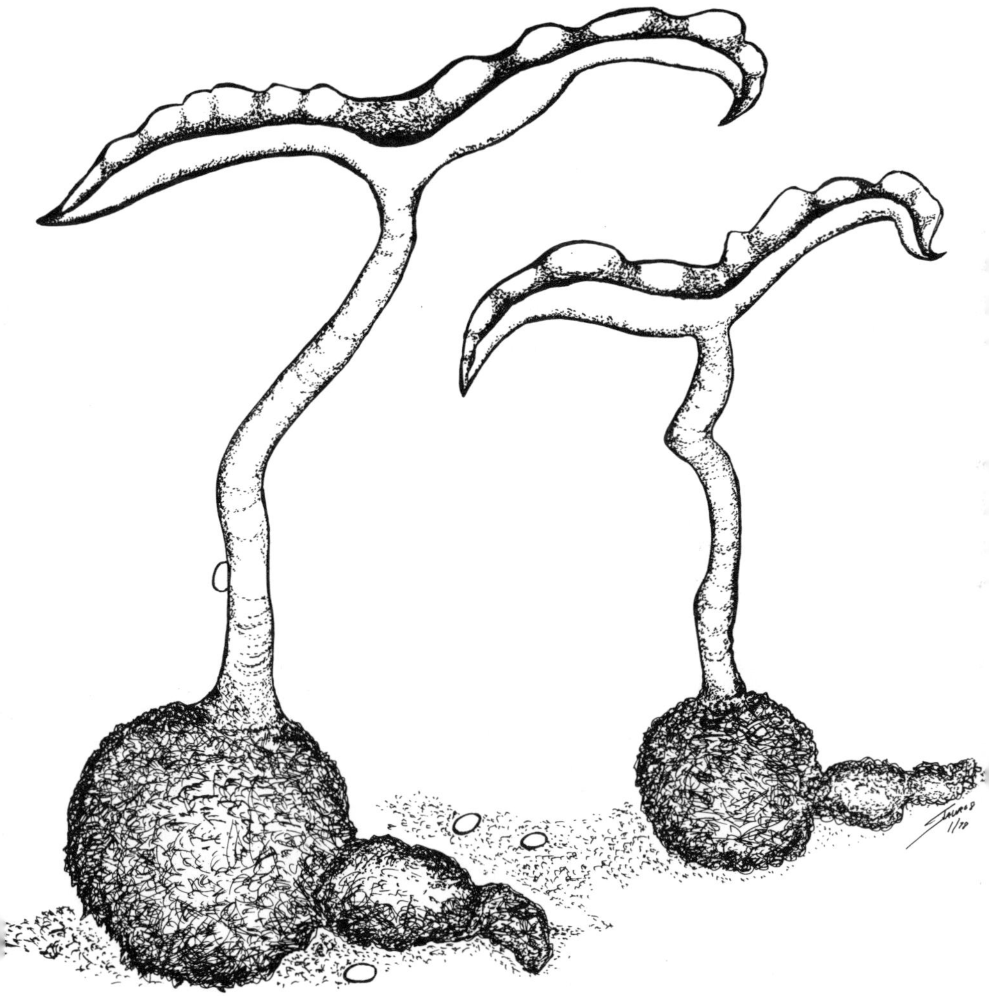

it may be adapted for countering malignancies. In both cases, bonellinin blocked further growth of the cells. If continued testing reveals that the substance itself is unsuitable for human use, it may still be important as a chemical model upon which synthetic cancer-killing compounds can be based.

In the final analysis, this is how the sea may prove most useful to pharmacologists—by providing examples of new compounds that can then be duplicated and perhaps improved upon in the laboratory. Marine animals may not in themselves be utilized as the raw materials out of which the new drugs are manufactured for commercial distribution. Harvesting these creatures from the ocean in the numbers that might be required could be disastrous to the environment. It has happened before that a sudden interest in a particular marine organism has led commercial collectors to sweep it from the sea in such massive quantities that the balance of life to which it contributed was irreparably destroyed. Culturing them in aquatic farms is a possibility, but right now we simply do not have the skill. However, we can realistically anticipate synthesizing on land whatever substances these animals have that may be of value to medical science, though reaching that level of knowledge can be a long and arduous process. At the University of Oklahoma, for example, Dr. Alfred J. Weinheimer has been working for six years to isolate the single pure active compound in an extract from the sea squirt that has scored the highest marks in experiments with cancerous mice. If left untreated, mice that have been injected with leukemia will die in ten days. Those mice given the sea squirt extract, however, have held out for as much as thirty days, which is as long as the test is allowed to run. Since some drugs now in use have kept the rodents alive for only twenty days or so, this extract is a product of enormous potential. Morever, it also suppresses the mice's capacity for forming antibodies and resisting skin grafts, making it a good candidate for assisting a body in accepting a transplanted organ.

It was Dr. M. Michael Sigel of the University of Miami who first noted these properties in material from the sea squirt, an animal that clings to rocks, growing either in colonies or as solitary individuals, and looks something like an enlarged grape. Specimens of it were brought into his laboratory as a result of a routine collection by divers under instructions to bring back for a general screening whatever was at hand. Those animals found in sufficient numbers to permit testing were then identified and put through a preliminary screening. This initial step in the development of a new drug is the one marine scientists have the greatest difficulty in getting funded. But there is some justification for this predicament. Right now there are literally hundreds of marine extracts with commercial drug potential in storage awaiting detailed analysis. They remain uninvestigated because there are not enough chemists and laboratories committed to this work. One reason they have shied away is that this next stage of the research is also not receiving adequate support. A frequent complaint of the scientists in this field is that there is too much pressure put upon them by some financial sources to achieve "quick results." When their work with a promising but enigmatic substance does not yield a dramatic breakthrough inside of a year or so, they too often find that their funder has lost interest and withdrawn from the project. Still, they have no doubt about their ability to master the chemical and biological mysteries before them; and considering how much has already been accomplished by the very small number of people and institutions devoted to marine biomedical research, their optimism does seem justified.

Where Onta is headed the ocean bottom was once dry land. At some ancient point along the scale of geologic time, the coast of the nearby mainland began subsiding. As the land lowered, the water swept in and turned it into the bed of a rising sea. Only the tallest peaks in the range of coastal mountains were not submerged, being transformed by the encroaching water line into a group of high granite islands. Shallow and warm, the new extension of the sea was fertile ground for the growth of reef-building corals. Isolated coral patches exist even in polar seas, but only where the ocean offers a temperature of at least 70° F. and sunlight can penetrate its depth will the polyps raise a reef. While the mountains were being reduced to islands, layer upon layer of coral skeletons were being laid around them and at the outermost edge of the sinking coastal shelf—millions of years of continuous life cast in the form of a barrier reef. Later in its evolution, the barrier was also fortified by the rising of a chain of coral cays. Unlike the high islands, the cays belong entirely to the reef, tracing their beginnings to the dead coral broken loose by the ocean and washed over the barrier's surface. After innumerable centuries of being ground into sand on the leeward side of the reef, this residue became a spit protruding a few feet above the highest level of the tide. Trapping vegetation that floated in with the currents and hoarding the droppings of each passing bird from the flocks of sea eagles, gannets, terns, herons, mutton birds, and frigate birds roosting on the barrier's shores, the mound of sand turned into fertile soil. Then the wind landed the seeds for a patch of grass and a few coconut palms, and somehow they survived amidst the terns nesting shoulder to shoulder in the sand. The cay had come to life. Beneath and around these now forested beaches, more than 300 coral species contribute their distinctive shapes and colors to the endless molding of the reef—anchorage for a spectacular array of animals, many of them bearing the gaudy appearance that blends best with the florid complexion of the coral.

Led by her instincts into this community, Onta touches bottom for the first time behind one of the gorgonian corals—a soft, branching specimen that sways in the currents like a young tree being tossed by an easy wind. Research with a gorgonian species from the Caribbean similar to the one Onta is huddled beside has revealed that the feathery coral possesses a sampling of the compounds known as *prostaglandins*—substances which are involved in the control of several vital life processes in mammals. Only minute amounts of these compounds are found in mammalian tissue, where they were initially singled out in the early 1930s. But the Caribbean coral yields its prostaglandin in a significantly higher quantity. Though the extract appeared at first to be inactive, chemists have since discovered how to convert it into two compounds that are biologically potent. In the opinion of Dr. Alfred J. Weinheimer, who discovered the presence of this hormonelike substance in the gorgonian, any one of the mammalian prostaglandins being studied at this time can be constructed from the material provided by the coral. Since the species involved is found in abundance throughout the Caribbean and is one of the faster-growing stocks, there seems to be no immediate danger of its being overcropped. As yet we do not know precisely how the prostaglandins affect the body's functions; nor can we explain why this complex biochemical material exists in an animal as primitive as the coral. But once again the means to help us preserve our health and improve our understanding of the basic chemistry of life has been given to us in generous amounts by the sea.

From an animal related to the coral by virtue of its also being a creature with stinging cells, the jellyfish

Aequorea aequorea, we have already gained one important diagnostic tool: a protein that can be triggered to glow in the presence of calcium or strontium. This bioluminescent substance, known as *aequorin*, makes it possible to detect calcium changes in humans that are often reflective of variances in the workings of our cells which may be signaling the onset of certain cancers and heart irregularities.

Outside the coral grove sheltering Onta, an animal shaped like a cylinder and studded with short, nubby bristles, is crossing the algae on tiny tube feet which are hidden beneath its body. It is a curious breed, the sea cucumber—so poisonous that fishermen can drive their catch into the open by rubbing the echinoderm's body along the rocks until it oozes those substances the fish must flee from or die; yet the slender pearlfish can dwell inside the sea cucumber's anal cavity unharmed. If severely agitated, the animal is capable of expelling every one of its internal organs and then retreating to a new location where it will eventually grow back everything that was lost.

Concerned at first with its powers of regeneration, Dr. Ross F. Nigrelli, now senior scientist at the Osborn Laboratories, went to Bimini in the early 1950s to study the Bahamian sea cucumber. In the course of his experiments with this creature, however, Nigrelli observed something that has long been known to native fishermen: any fish in the vicinity of the sea cucumber when it disgorges its viscera will die. Even when the water from Nigrelli's tank of sea cucumbers was diluted through two other tanks before it reached the one containing the fish, they succumbed. His attention directed then to the visceral mass itself, rather than the animal's ability to regenerate it, Nigrelli was able to isolate its active agent, a chemical factor which he named *holothurin.* Further experimentation showed that one fraction of the mixture acts upon muscle; when applied directly, it causes the muscle to contract. This property makes holothurin useful in physiological studies. But an even more significant discovery was waiting to be made. In every test, Dr. Nigrelli was observing that the drug was also having an arresting effect on essential metabolic processes. Guided by this evidence, he began testing it as a potential anti-cancer agent. Injected into cancerous mice, holothurin reduced the size of their tumors and added days to their lives.

While work with this substance continues, investigators at the Osborn Laboratories have also been searching for related materials in other echinoderms collected from both the Atlantic and Pacific Oceans—an effort that has so far uncovered several substances with promising pharmacological effects. And Japanese scientists, working with another species of sea cucumber, have extracted a mixture similar to holothurin that is highly active against fungus infections. There are many echinoderms that can grow replacement parts for their anatomy. Cut an arm from some types of sea stars, for example, and

Sea cucumber expelling its internal organs

not only will they form another to take its place, but the severed limb will itself develop into a complete animal. Whether in the field of regeneration or pharmacology, this group of animals has much to offer that may help improve our own physical condition.

Feeling exposed and vulnerable as tiny swimmers weave about her through the branches of the treelike coral, Onta begins searching her immediate surroundings for a more secure haven. This is her first prolonged look at the bottom world, and as she probes it for a cubbyhole of her own she is also sensing the habits of her new neighbors, recording their ways. In a clearing beside the coral forest, a number of small and especially colorful fish have settled into conspicuous positions in the open. Contrary to the ordinary behavior of animals their size, they are eager for the arrival of larger predators in their territory. When one does enter, it immediately becomes calm and stationary, and the little fish is allowed to scamper about its body, nibbling it clean of parasites and sore flesh. Often there is a line of creatures waiting beside one of these cleaning stations to be hygienically groomed, their wars with each other temporarily suspended while they are being treated. But Onta also notices that a butterfly fish visiting a sabre-toothed blenny, which is patterned as a cleaner and wriggling in the same manner has actually been led into a trap. Once it is relaxed and off guard, the butterfly is suddenly stabbed by its host's teeth, bitten into, and before it has recovered and is ready to fight back the imposter is gone with a piece of healthy flesh. Instinctively Onta understands this kind of deception, being a creature of similar guile whose own masquerades will becomes more pronounced as she grows older.

III. Night

Slowly the light of day begins to lift from the deeper reaches of the sea as though its dangled colors were being reeled in by the departing sun. Gone first is the scattering of violet rays that gave barest illumination to those creatures dwelling below a thousand feet—inhabitants of a world whose brightest hour is but a slighter shade of abysmal night. Swimming in a flow as black as time beginning, many animals at this depth carry inside them either glandular cells or pockets of bacteria that respond to an impulse for light with sharp, luminous flashes, their cold sparkle being caught by eyes that have grown giant and bulging from aeons of straining to see where other earthly life would be blind. In the zone just above them, the water this afternoon had been as vibrantly blue as the sky at twilight in that moment before the first star is visible and the horizon still holds a pale remembrance of the fallen day; now it too is blackening. Next to fade as the sun draws closer to the rearing waves will be the green hues that spread as far down as 500 feet and are the ones most easily perceived by fish; to be followed in quick succession by the red rays that warm and light the uppermost layer of the sea. Darkness is rising early this evening over the mid-ocean cliffs and mountains as the white caps bubbling on the surface trap and reflect most of the sun's lowering glow. While it still has fire enough to blaze a golden trail across the spilling swells, the currents below will be immersed in deepest night.

An hour before sunset, the darkness scaling the ocean prompts numerous fish grazing near the surface to begin drifting down toward the reef where Onta lies hidden in her cave. Leading the descent are tiny wrasses that wriggle between rocks and under ledges, finally resting on their sides in clefts too cramped for an enemy to enter or partially burrowing into the bottom sand. Behind them in the plunge for shelter come other plankton-feeders from the water column above the reef. Within each species it is the smallest individuals that retire first, their brilliant daytime markings fading as they hover about the coral bluffs so they can tuck more securely into its crevices. Inside the stony recesses of the reef a parrotfish is releasing through its blue-green skin diaphanous threads of mucus that close around its body like a blanket. With minutes remaining until the sun is completely extinguished in the sea, a procession of gobies, each one smaller than a human finger, approaches the whorling passageways and corridors of a massive brain coral seeking harbor for the night beneath its delicate polyps. During the day, the polyps stayed tamely withdrawn in their limestone cups, avoiding all light; but in the dark they become irascible carnivores. Any crustaceans small enough to be seized that errantly brush against a polyp's skin are numbed by the toxic spears fired from its nematocysts and then flicked mouthward by the sweep of its tentacles. As the gobies settle down for the night, the coral polyps are just awakening, their stinging cells extended above the fish like an arsenal of protective missiles.

For about thirty minutes after the flight of the sun is over many of the fish returning to the reef—most noticeably the pearly damsels, the surgeons with their bladed fins, and the radiant butterflies—hold in almost stationary patterns beside the portals of their evening cover. Coming out to join them are those species that rise in shadows for the hunt—the inch-long cardinalfish, whose vermilion cloak will wash to silver-gray with a single black band when it patrols the milky branches of the staghorn coral, the larger squirrelfish, and the bigeyes and soapfish, which do not swim far from their beds to feed, mustering in advance of the others. Below this milling cloud, the grunts which had passed the day schooling lazily near the bottom are fanning out to search the neighboring groves of seaweed and soft, flowering corals for the mollusks and crustaceans that inhabit the sand. Grouped together, they had provided a measure of safety for each other, their tangled movements making it harder for a predator to concentrate its hunger on any one of them. But now, as the grunts range individually away from the reef, they are exposed and vulnerable even though the bright yellow stripes they wore while assembled are gradually being covered by drab blotches.

At this hour not only the fish in dispersing schools are being threatened. Many of those pausing beside the openings to their coral compartments are also caught shedding their noontide defenses while they are not yet fully adapted to the dark. Though their quiet congregation seems to say that they are in the midst of a period of comparative safety, they are actually in grave danger. About the rocks are predators who rely on this moment to fill their stomachs and are gliding in for the kill, their victims being taken as easily as if they were bait on a hookless line. They do not have long to feed, however—ten, perhaps fifteen minutes at the most. Then, like a spray of buckshot, countless fish and bottom-crawling creatures pour out of the corners and alleyways of the reef and struggle up from their tunnels in the sand to take their turn in open water—among them, the shrimps and crabs and lobsters that Onta favors in her diet. As they flood the reef, the last of the fish that have been lingering about like children stealing extra seconds for their play before being forced to sleep dive suddenly for cover, signaling the end of the early evening calm. Instantly, the night hunt begins.

Crawling from her lair in the craggy hollows of the reef, Onta is alive in a way repugnant to most humans. Boneless and globular, her torso has no appendages—no fins or claws or flippers—giving identity to its shape. When this sleek, rounded body floats higher than her flattened head and rises behind her black and yellow eyes, as it often does, it looks more like a deformed cranium—the bald and swollen brow of a monster intelligence—than the mass of internal organs it really is. The limbs Onta has are long and serpentine and grow out of her head like the trunk of an elephant. In all there are eight, webbed to one another by membranes that circle them at their base. Most of the time she uses them as arms, and they flex and curl and slither through the water, swarming after things to be seized and eaten; but they are also her legs. Each is covered on its underside by smooth white disks—hundreds of them flowing in two rows like a field of mushrooms breathing in the wind. These are Onta's fingers, and the source of her strength. With the enormous suction of these disks she can grab and hold, and not even the power of several men pulling in unison can make her let go. Until she wills otherwise, the disks probe over the surfaces of things without sticking; but feeling something she desires, Onta tenses muscles that hoist up their centers, creating in each a vacuum that clamps its muscular rim in place. Then whatever has been touched is possessed.

All day Onta has been huddled between the rocks in a space barely bigger than herself. Shorn by evolution of the hard, protective shell that armors other mollusks such as clams, oysters, and snails, the octopus has developed a compulsive reluctance to be seen and feels menaced by unobstructed places. So great is its instinct for solitude that it finds comfort in caves whose walls shield it as closely as a second skin. But at night the need for wandering is also in its mood. Once the sea had thickened to the hue of a single concealing shadow, Onta's hunger began gnawing at her fear of exposure. Later there will be a half-moon and just a few clouds too thin and scattered to keep its frosted beams from penetrating to the reef and revealing her path. But she has no sense of that now—only that the darkness she treasures will be diluted by the colorful sparkle of phosphorescent animals, and to this she is accustomed. Though still afraid of the creatures outside that might pursue her, she is driven by the stirrings of her appetite to leave her den. Like water seeping from a jar imperfectly sealed, she

A lobster is caught

squeezes through an almost imperceptible doorway, her arms twisting out first to assure her that nothing dangerous will be met this soon.

Moving along the bottom, Onta walks by casting her rubbery limbs on ahead and reeling herself forward to where their suckers are grasping the ground. Into every crevice, every hole, slides the feathery tip of an arm searching gingerly for a clam or a scallop or some other small prey easily snatched. Though clams and scallops are among the simplest mollusks, possessing neither head nor jaws nor teeth, they are not wholly defenseless. Like knights dressed in armor, they shield their soft, vulnerable bodies in a bivalve shell which is cast from the secretions of their mantles. From within this skeletal fortress, a bivalve such as the scallop keeps contact with the outer world by means of long, brightly-colored tentacles that fringe its mantle, which also houses as many as 120 tiny eyes that peer out when the shell is slightly ajar, on guard against the approach of creatures like Onta. When one is sighted, the scallop erupts in flight by rapidly opening and shutting the two halves of its shell. A clam being pursued seeks its safety in another direction. Slipping a hatchet-shaped foot out between the lips of its shell, it burrows deeper into the mud. This retractable appendage is also used by some bivalves to creep across the bottom sand. In a few species, it is so strong that it can actually bore through granite.

Above and around the octopus as she prowls is the tinsel glint of glass-like worms and jellyfish; where certain one-celled dinoflagellates have clustered there is a green, fairy-lantern glow. Alarmed by Onta's presence, a brittle sea star which has been feeding amidst the glittering plankton scuttles away, its nervousness shown by the lights dancing up and down its arms. Traced by the water's luminescence, the octopus slips into disguise. Colors imitating the coral branches and mottled sea crags being crossed pulse through Onta's flesh, masking her as she forages. The paint of a million tiny pigment cells enables her to coat herself in any pattern she sees. Triggered by her eyes, these muscle-controlled cells expand, spreading puddles of color through her transparent skin. Other mechanisms make it possible for the texture of her skin to change, and on those occasions when it loses its customary smoothness and bristles with ridges and wart-like protrusions, Onta looks even more a part of her surroundings.

As the octopus slides cautiously toward a patch of turtle weed growing beside the reef, she is watched with tense concern by several crabs whose flickering antennae and telescope eyes are all that might be visible to her from under their own camouflage. Three days ago, one of the crabs, using its pincer claws as scissors, had torn from a sheet of sponge encrusting the coral a piece large enough to drape over its shell. Meticulously it had worked at trimming the hem of its new dress; now, held in place by the animal's two hind legs, the cutting covers it perfectly. Since the sponge feeds directly on the sea, absorbing all the nutrients and food substances it needs from the streams of water filtered through its body walls, it will remain alive while carried by the crab; and because the sponge is hunted only by a few snails and fish, it will not lure creatures like Onta who would make a meal of the arthropod crouched beneath it.

Amidst the dissonant sounds and ceaseless motion of life in the oceans, the sponge is one creature forever silent and sedentary—a sightless witness to the hunting and seizing of prey in which it takes no voluntary part. Wherever it is born it remains: a mass of cells so primitive in nature that no organs can be detected inside its structural cavities and it seems more like a plant than the animal we know it to be. Other creatures, such as coral, live out their time fixed in place and have been arranged by evolution in flowery shapes; but they were conceived to use their stationary position as an ambush. The sponge attacks nothing. Mixed in with the water it ingests are bacteria and other microorganisms that dwell inside its canals yet seldom cause it harm. Exactly how long it lives, no one can say. Some specimens have been kept under fairly constant observation for about thirty years and shown no change. Considering our own span, the sponge seems virtually ageless. Like an ornament placed with no other purpose but the sea's decoration, it thrives in more than 5,000 species and a multitude of shapes and colors at almost all depths throughout the world's oceans. Viewed next to creatures that swim with extraordinary grace and possess startling strength or a

Collecting sponges

high order of intelligence, the sponge may seem uncommonly dull, hardly worth noticing. But in nature's terms it is a marvel of biological success. In a world where only the very largest beasts and the most efficient killers are usually granted long life, the sponge endures as few others can. Though lacking an obvious means of defense, it is ordinarily as safe from the dangers of its environment as if it were sealed in glass.

It was an accidental discovery made more than two decades ago during the study of a sponge from the Caribbean that provided the starting point for the eventual synthesis of an important anticancer agent—*Ara-C*, a drug that helps reduce and destroy malignant tumors and was finally released to physicians in 1969. While treating a specimen of the sponge *Tethya crypta* with an organic solvent to determine its fat content, Dr. Werner Bergmann and his students at Yale University noticed a strange white substance forming in the flask. Puzzled by the appearance of this material at a stage when they hadn't expected any crystallization, they took up its analysis. What they had stumbled on were some previously unknown members of the class of compounds called nucleosides. These are the basic building blocks of the nucleic acids—the RNA and DNA that carry the codes governing the growth and reproduction of animal cells.

Not only had these particular compounds never before been found in a natural state, but they were so unusual in structure that Bergmann and his assistants were unable to explain their existence. Even today, more than a quarter century after their discovery, we still do not know how or why they are manufactured by the sponge. Nevertheless, Bergmann's identification of these arabinosyl nucleosides had given cancer researchers a promising direction to explore. Within nine years, other scientists had succeeded in synthesizing a related compound, *cytosine arabinoside*, whose potency was first demonstrated in 1961 in the treatment of leukemia in mice. After eight more years of laboratory testing, the new medicine was ready to be used in the saving of human lives. Relied upon extensively today, Ara-C is also the focal point of research being conducted in Japan to produce a sister drug that will be even more effective. Using cytosine arabinoside in combination with fatty acids, Dr. Hidehiko Kobayashi and Dr. Torao Ishida of Asahi Chemical Industries have fashioned a new series of nucleic acid derivatives. In early tests with mice some of these advanced compounds have shown themselves to be more than four times as active as Ara-C against leukemia.

In a similar development, French scientists screening an extract from this particular Caribbean sponge for possible effectiveness against cancer were surprised to learn that the substance in question was clearing their test materials of all viral contamination. Subsequently, research teams at several American laboratories and the pharmaceutical firm of Parke-Davis & Company were able to synthesize a new compound, *adenine arabinoside*, which is the first antiviral agent that can be taken internally. This is especially significant because viruses operate by penetrating the body's cells and then reproducing *within* their walls. For a drug to be successful against them, it must be capable of opening a way for itself through the cell wall without also destroying the cell in the process. Though *Ara-A*, the latest drug based on the research initiated by Bergmann, is only effective against herpes encephalitis—a virus that attacks brain cells and usually results in death—and the eye infections caused by another herpes strain, it represents an important breakthrough in the chemical treatment of disease. Finally there is one antiviral drug (of three licensed by the Food and Drug Administration) that protects the integrity of the cell while countering the virus within it and we can reasonably hope others will follow, however slow their advance may be.

From these and other discoveries it has been established that sponges can be an excellent source of health-restoring materials, particularly antibiotics. Nearly all the sponge species tested to date have yielded numerous substances with strong physiological properties. But an enormous amount of basic research remains to be done before other medicines derived from sponges become available. Since very little is understood about the metabolism of the sponge, the biologist cannot always be certain that the activity he or she is monitoring is attributable to the animal itself. It might also be caused by the bacteria and algae that frequently flourish in the creature's many canals. Numerous forms of algae are known to be antibiotic; also, some bacteria protect themselves by emitting substances into their environment that will kill off any invading microbes. Once it is possible to culture sponges independently of their accompanying microorganisms this issue will be decided. Until then, the assumption is made that whenever a large amount of active material is obtained, the sponge is producing it; and the painstaking task of fixing the exact chemical composition of these highly active substances continues.

Clinging to the turtle weed is another decorated crab, this one having hooked clumps of algae and the fragments of a sea fan onto the hairs sprouting from its back, its plumage so carefully arranged that to Onta the animal seems to be but one more plant in the garden she is skirting. A tiny, swimming crab whose color matches the waving grass exactly also escapes detection. Not all of the crabs in Onta's track are as lucky. A young female pea crab which has nestled inside a living clam where she sustains herself by scraping food from the gills of her host is too dependent on the bivalve's survival for her own. When Onta scoops the clam out of the sand and cracks its shell, she too is swallowed. Pleased by this unexpected addition to her small catch, Onta perches atop a coral mound and blooms the rippling colors of her contentment. But in the sea relaxation is the supreme luxury and never lasts very long. To the left of the octopus there is a sudden swiftness in the water that does violence to her senses and destroys the moment of her rest.

From the open range beyond the reef, a hammerhead shark has traveled here in search of the small, bony fish that cloud the night and has been surprised by the sighting of a more appetizing prey, a round stingray that is cruising the bottom not far from where Onta is sitting. Veering sharply for the dishlike ray, this wolfish invader has angled himself in the octopus's direction and it seems to her that she is his intended victim. In the instant of this perception she flees. Just below her head there is a parting in the mantle of skin and muscle that girdles her body, and jutting out from this opening is a funnel through which water is expelled as she breathes. Forced to race for her life, it is the action of her funnel that drives her on, not the motion of her eight arms. In through the mantle cavity water is drawn more rapidly than usual and shot from the funnel in powerful blasts that thrust Onta through the sea. But the shark has been attracted to the swirling wake she is leaving and changes course again, this time to charge her. Panicked by the speed with which he is approaching, she squirts a thick black liquid through her funnel, a puff of ink that seems to resemble her in shape. Immediately, as this veil appears, Onta pales and jets away in the opposite direction. No longer a thing of startling rainbow flashes, she is ghostly, an almost transparent white as she tries to escape. But this initial screening spurt has failed to distract and deceive her pursuer, and seconds later she fires another, once more swerving behind it to elude the onrushing shark; and still other bursts are released in a

Fleeing a hammerhead shark

sequence so rapid it appears as though she is being drained of her blood. Confused by her murky decoys, each one offering up yet another vaporish target, the shark cannot decide which way to turn next. This hesitation is fatal to his attack. By the time the inky forms have dissipated, Onta is gone and the shark is left looking in vain for the stingray which had first excited him but now has also vanished, its lantern eyes prying elsewhere for a meal of shrimps and crabs.

Brutish in appearance and under the command of a rampaging appetite that makes virtually any sizable life in its path a potential victim, the shark awakens in us prehistoric fears of being cornered and consumed by the sharp-toothed giants of the animal world. Ashore we laugh at its diminutive brain and its wooden gaze; but once at sea we can be terrorized by the simple mention of its name and given nightmares by a moment's thought of the indiscriminate savagery of its feeding frenzies. Granted a single wish before landing again, too many of us would cheerfully condemn this animal to extinction; few indeed would imagine that the same primitiveness we find so fearsome in its outward behavior also endows it with a biochemical capacity to resist illnesses that destroy large numbers of us daily, and in this sense it is to our great advantage that a beast like the shark does exist.

Just as terrestrial animals are susceptible to a host of deadly diseases, so are the oceans' creatures plagued by virus infections, fungi, parasites, ulcers, tuberculosis, cancers, and many other critical disorders. Several years ago, for example, yellowtail flounders schooling off Nova Scotia were stricken with a devastating ailment. Canadian authorities immediately shipped specimens of the fish to the Osborn Laboratories so that Dr. Ross F. Nigrelli could identify the disease. His investigation revealed that the flounders were suffering from a fungus which infects every organ of the body and is known to have flared up in other species, such as herring. This time, however, the outbreak was epizootic—restricted to this particular fish and only those living within a certain sector of the sea.

What remains to be deciphered are the reasons why a given disease erupts in the aquasphere. Is it the result of infectious agents, hormonal or nutritional imbalances, hereditary influences, or is it simply the inevitable consequences of old age? Might it be initiated or heightened by an unexpected bloom of planktonic organisms, or is it in some way related to the increasing amount of pesticides, industrial pollutants, and radioactive fallout and wastes awash in the sea? Understanding the disease processes at large offshore is an essential prerequisite to any attempt at fish farming on a grand scale. If enormous quantities of fish are to be bred in confined areas without mass mortalities, the sea farmer must be able to anticipate, recognize, and treat whatever illnesses threaten his stock.

Marine biologists have been aware for some time that sharks seemingly never contract virus infections or develop cancerous tumors. At the University of Miami, Dr. Michael Sigel and his co-workers reasoned that the primary explanation for this remarkable occurrence would most likely be found in the physiological character of the fish rather than conditions peculiar to its waters. As indicated by their initial studies, this was indeed the case, the sharks' immunity traceable to the presence in their blood of a single protein, IgM. At first, however, it was difficult for them to accept their own answer. Though IgM belongs to the group of proteins characterized as immunoglobulins, the fabric from which antibodies are made, it is such a primitive one that the investigators could not believe that it would, by itself, render these animals impervious to the intricate disease patterns of the virus. In humans, for example, IgM appears mainly during the fetal stage and as the infant grows is subsequently superseded by more complex immunoglobulin forms—the G, A, and D proteins. But they were soon to be given an astounding underwater demonstration of IgM's unique properties.

While conducting operations on sharks which involved removing some of their internal organs and then returning the animals to their offshore pens for observation an accident occurred that should have resulted in their death, but in fact did not even weaken them. As it happened, the sea water corroded the sutures holding the surgical wounds closed before they were healed. Now each shark was swimming about with its intestine poking through the gash in its belly; but, oddly enough, not one suffered the infection that would have been inevitable—and fatal—in any animal of a different chemistry.

Faced with this evidence of its natural effectiveness, the experimenters began testing the IgM protein from the shark's blood against selected viruses and carcinogens in mice. True to its chemical code, it killed both the standard infections and the cancer cells—but too imperfectly to be considered the next miracle cure. Because it is an antigen, and as such stimulates an organism injected with it to produce antibodies that will neutralize its effects, it is most successful when a tumor is small and therefore more easily eliminated. Given a second dose of IgM protein, a mouse's body in essence sides with the cancer that is killing it by laboring to eradicate this foreign substance despite its medicinal qualities.

How this unusual mechanism, or the knowledge gained from its study, will eventually be incorporated into a prescription for renewed health remains to be seen. But considering its source, we would do well to amend our attitudes toward animals like the shark whose natural behavior we deem odious. Each year, on the average, there are less than thirty documented cases of sharks having ravaged human beings. This year alone more than 400,000 will die of cancer in the United States. At what cost to ourselves would we be banishing these creatures from their own domain, a part of this earth we humans cross but do not inhabit?

Backwards Onta swims, guarding her retreat from the attacking shark with eyes that have the same wondrous detail as those of a human and may actually enable her to see more clearly through the watery curtains of her world than we do in the brighter light on land. This is due to the position of the ganglion cells in relation to those of the retina. In the octopus, they are behind the light-gathering retinal cells and therefore do not interfere with incoming images; but in our optical system they lie between the retina and the lens, obscuring vision except in the macula, the one small area of the retina that is unobstructed.

Not until Onta is some distance from where the hammerhead was last seen does she dare spread her arms slightly, opening the membrane between them, and float down to the ocean floor. Though her hunger is returning as she slowly calms, she will not reveal herself again this evening—especially now, with the moon's silver glow drizzling down around her like an incandescent mist. Curling beneath a shelf of rock, she waits patiently for her food to scurry by. To the natural shape of her new fortress, Onta adds a heap of sand and stones that walls her in more completely. From a broken clam shell being scooped into this mounting palisade a dozing wrasse is tumbled, much to her delight. Uncertain of its heading, the startled fish slams into one of Onta's arms and is pinned by a sucker. Quickly she snakes the wrasse back to her mouth, baring the black horny beak that is otherwise hidden inside the pulpy flesh at the base of her arms. Encouraged by this catch, small as it is, Onta concentrates on improving her concealment; she is not so much cowering from her enemies now as she is laying a trap.

Awakened prematurely by the filtering moonlight, other diminutive wrasses are voluntarily quitting their resting places. For a few seconds before swimming off they widen their mouths as though yawning and limber their bodies with a series of contortions that shake out the last vestiges of their sleep. A crab scuttling toward their chorus line disperses them; but it is not after fish. Overtaking the sea anemone it has spied hitching a ride on the shell of a large snail, the crab transfers the coral-like animal onto its own back. Shielded by the anemone's waving tentacles, which are spiked with stinging cells, the crab is now free to scout the bottom for food without fear of being pounced on from above by a hungry octopus.

All this Onta witnesses without betraying her position. Unless the water she is breathing fills with the scent of an enemy, she is prepared to remain motionless for hours in order to take her prize. Only by the pulsing of her suckers and the occasional quiver of a tentacle could a creature peering this way tell that she is alive. But none so far detect her. Onta has the advantage of a brain more highly developed than that of any other boneless animal, and it is behind this intelligence that she ultimately hides. Though well protected by the rampart she has just built, she makes herself feel even more secure by extending her arms and drawing the membrane between them across most of the space left open at the top. Behind this veil she eases from her mind the thought of things that might linger outside and with their staring disturb her ambush.

Thousands of years of storytelling, sworn to and believed by people who imagine the worst of whatever is alien and unknown, have condemned the octopus as a "devil fish"—a killer of humans. With scientific pretense, this ancient misconception has been exaggerated by writers such as Victor Hugo, whose hatred of this creature induced him to describe it as a fiendish being that would consume a man trapped in its cave under the sea.

"No grasp equals the embrace of the octopus. It is the pneumatic machine attacking you. . . . . Your muscles swell, your fibers writhe, your skin cracks under the foul weight, your blood spurts forth and mingles frightfully with the lymph of the mollusk. The creature superimposes itself upon you by a thousand mouths. . . . The tiger can only devour you; the octopus, oh, horror! breathes you in. It draws you to it, and into it, and bound, ensnarled, powerless, you feel yourself slowly emptied into that frightful sack, which is a monster," Hugo lectured the world in his novel, *The Toilers of the Sea.*

In reality, however, the octopus is a timid creature. Old drawings may picture the beast in giant form malevolently wrapping itself around vessels under sail and tearing down their masts, but actual experience with different species of this invertebrate has demonstrated that the greater the size of the octopus, the less likely it is to venture out against even a single person. Though the smaller octopuses are generally more aggressive than their larger brethren, they too shrink from the diver in their realm, either swimming off at first sight or sinking deeper into the crannies of their homes. The claim made by one man from Australia during World War I that an octopus charged up from the sea and tackled him, its tentacles grappling his waist, arm, and legs is, like Hugo's account, exciting to read but hardly indicative of what can normally be expected from this animal.

In spite of its innate shyness, however, the octopus cannot be considered entirely harmless, as one man learned most painfully while exploring the surface waters off the coast of Chile. At a depth of about 20 feet, he was suddenly struck in the groin by a blackish fluid shot at him by an animal he was unable to see well enough to identify. After leaving the water, he is reported to have experienced a burning sensation in his genitals, and the following day the skin there turned dark purple and his discomfort became agonizing. Since no other animal in the region is known to possess this kind of inklike substance, it is believed that the diver (who recovered completely from the encounter) had inadvertently come too close to an octopus and was unfortunately hit in a very sensitive area by the creature's defensive spray.

Of far greater danger, though, is the animal's bite. The saliva it drips into its prey contains a mixture of compounds that are highly toxic, even to humans. People imprudently handling the octopus and causing it to lash out with its beak find their bodies going numb where they have been struck, may become giddy and vomit, and for a time be partially paralyzed. In the Pacific and Indian Oceans there is a species, *Hapalocheana maculosa*, that fully grown measures just 4 inches across with its tentacles spread but has in its glands the deadliest venom of them all. Favoring tidal estuaries and shallow inlets where it takes shelter from the sun under cool rocks and inside vacant shells, it is distinguished by its size and the blazing blue rings that light around its body when it is disturbed. Unfortunately, its fondness for rock pools places it within easy reach of children and nature enthusiasts, and its lilliputian stature entices them to hold it. Reduce the octopus to doll-like dimensions and splash it lavishly with color and it is no longer ugly in our eyes. Somewhere in the course of having elevated ourselves out of the animal world we seem to have acquired the notion that if a thing be small and pretty it must also be harmless. As regards this animal, at least, nothing could be further from the truth. For two individuals known to science, this quixotic view of nature has resulted in death; several others have been bitten and recovered, but while in the throes of the toxin suffered horrendously.

Within minutes of being pierced by the pigmy beak, the victim's lips and mouth begin to tingle, then all sensation is drained from them. A chilling numbness spreads through the back of his head and across the neck, and as it does his speech slows and thickens and his vision blurs. Immediately his arms weaken and he can no longer stand. Only with great effort is he able to force his lungs to swallow air. For a while he vomits and quivers convulsively, but this sense of movement is misleading. Every muscle in his body is being relaxed by the poison, quieted to the point of going completely limp. Soon he will not even be able to blink his eyelids. Without artificial respiration, it is almost certain that he will not last the four to eight hours it takes for the poison to fade.

Though the salivary toxins of the octopus have been under study since the early part of this century, we still cannot offer conclusions as to how most of them function chemically. But from one, found only in a Mediterranean species first captured in the Bay of Naples, a compound known as *eledoisin* has been taken which is a powerful hypertensive agent.

It is ironic that the smallest of the octopuses, the one that does not repel us, is the one we have good reason to avoid. Nature has a way of reminding us from time to time that we isolate ourselves in ignorance when we judge her creations by how well they conform to our self-styled definitions of beauty. Seen by the uninitiated, a large octopus is something to be feared and backed away from, even when it is in an aquarium, behind glass—a creature cold and slimy, so repulsively different from ourselves that they perceive in the shape and motion of its life the thing they call evil. Is it because the creature's eyes have the gleam of high intelligence, the expressive stare that hints at an ability to interpret and decide? We have never really lived comfortably beside animals we could not rule as absolute masters, and it is those that glance back at us with the least sense of cunning that we most prefer to watch. If, as philosophers have said, the eye is the window to the soul, might we not then despise the octopus because its reflective gaze seems to challenge our own?

A crab coming within striking distance of Onta now will be taken not with a leap but by one or two arms slowly unwinding over the rock in its direction. Easily overpowered, the crab will be paralyzed by the poison in Onta's saliva, cracked open, torn apart, and devoured; then its empty shell will be used to reinforce her barricade. But before this can happen, there is another disturbance in the water, a drifting smell that drives out her hunger and causes her to flush with fear. Moving closer to the octopus's lair and poking its snout into every sizeable hole in the reef along the way is the creature she dreads most—the moray eel. For millions of years, these large eels have sought out Onta's kind, and in that time they have become adept at battling each other. This particular one is bigger than most, being nearly 10 feet in length, and can swallow a small octopus whole with its foot-long mouth. Stabbed by the eel's sharp, grasping teeth and drawn steadily back toward its throat, the octopus may try to force its release by squeezing the moray's head with its tentacles; but this act of desperation is easily countered. Circling itself with its tail so that it forms a loose knot, the serpentine fish will then pull its head backwards through the loop until the tentacles strike against its muscular body and are brushed off. All the while the eel will continue sucking its helpless prey deeper into its throat.

While the sea was sunlit, the giant moray kept itself concealed in the rocks, occasionally letting its head slip into the open only to be taunted then by the small fish it would easily snatch from sleep and consume later that night. An angelfish approaching the cloistered eel would suddenly swing to the side, flashing its fins near the razor-toothed mouth, which can crush human bone with its bite; a tang, even more willing to risk capture, darted forward and nipped at the moray's head. A creature of poor vision conditioned to coming out only after dark, the eel responded to these assaults by coiling down in search of deeper cover. During its first forays this evening, the moray nosed into the nests of those parrotfish that cloak themselves with a mucus which holds in their scent; and although it bumped them with its snout,

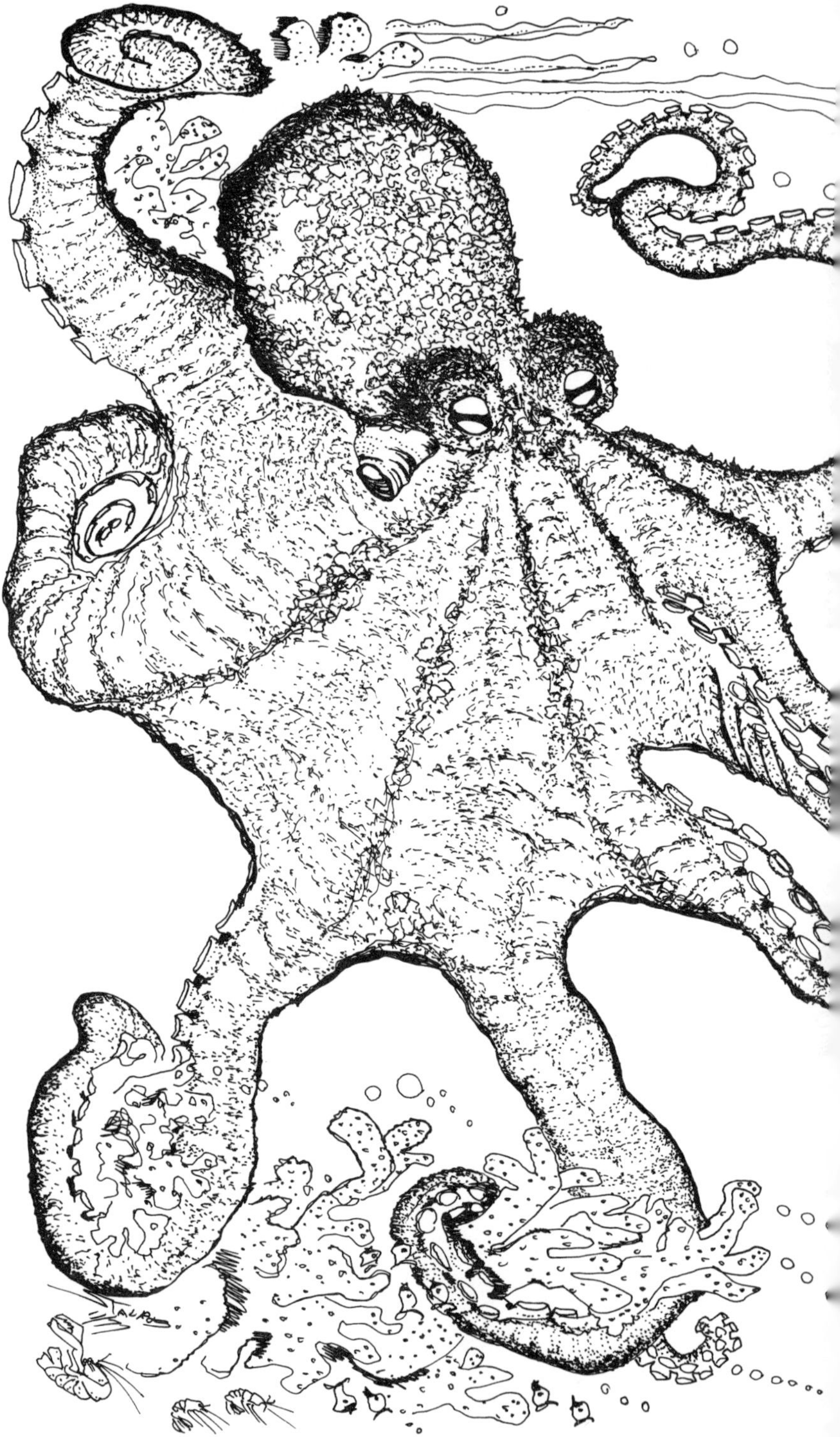

Attacked by a moray eel

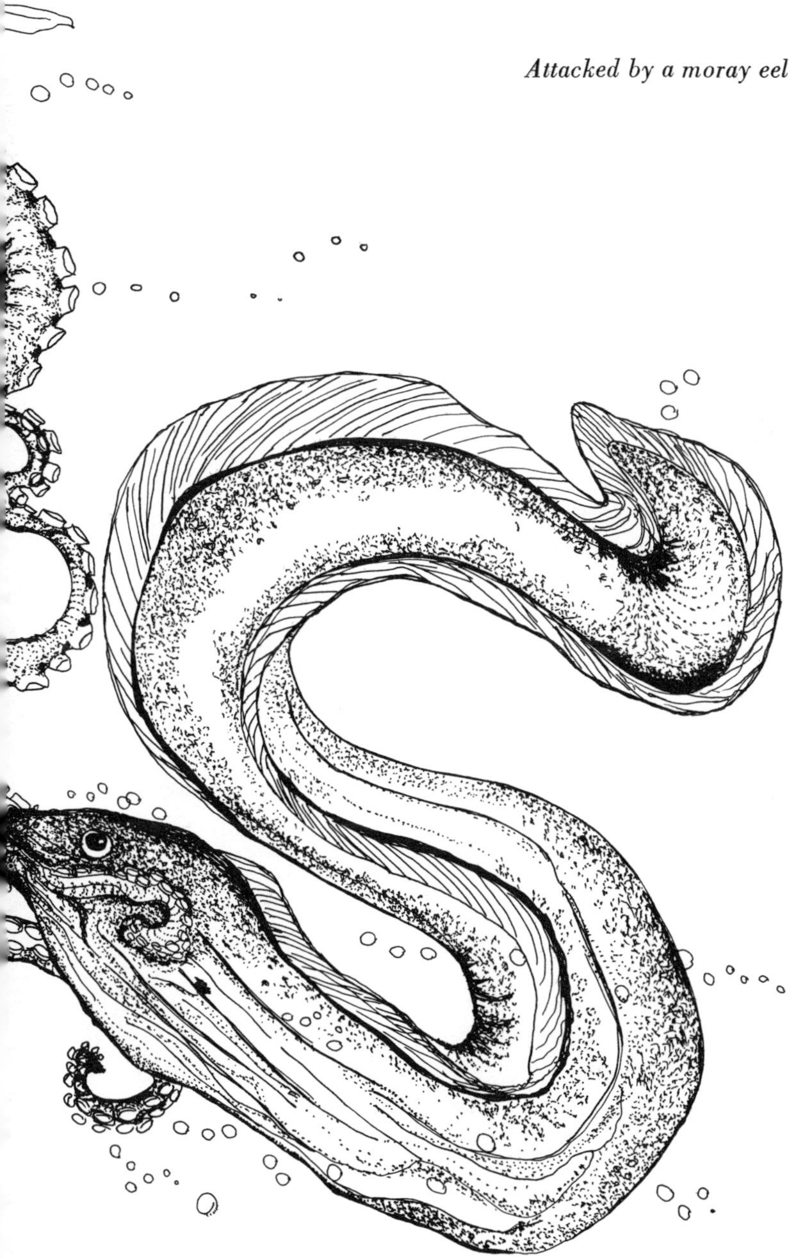

it was unable to sense them as prey. Its hunger sharpened by these frustrations, the eel has picked up the chemical track of its most ancient enemy—a faint smell of mollusk that is leading it to Onta.

Alerted to the danger by her own olfactory sensors, Onta turns herself into a near-perfect replica of the walls of her cave and braces for the eel's attack. Should the moray fail to locate her, she will let it pass unchallenged, though once before she killed an eel only slightly smaller than this one and ate its head. While she ranks among the larger of her breed, measuring 16 feet from the tip of one arm to the tip of its opposite member, she relies on her size to enhance her chances for survival in the event she is cornered and made to fight. Truly defensive by nature, she has no urge for combat with this creature. Unless the eel ventures into such close view that their collision is inevitable, she will not even fire her ink, however threatening its scent. Aeons of fending off the snapping charges of the moray and conger eels have patterned in the octopus an instinct for withholding its sepia spray until the slithering fish is almost upon her and the masking cloud will strike its snout while still fully concentrated. Once dissipated by the surrounding water, the cephalopod's ink will merely stimulate the eel's appetite, goading the creature into butting forward again; but if the purplish stain lands on target before it has been diluted, it will anesthetize the eel's olfactory organs and the moray will then crawl harmlessly over the octopus as though she were indeed a bit of rock.

With the same sureness of aim with which she would volley water at her spawn to keep it from being fouled, Onta points her funnel toward the mouth of the cave like a cannon, ready to fire a numbing salvo at the eel when it peers in. But the moray has snaked up to the entrance alongside the rock, where she cannot see it; and instead of slipping its head in cautiously, it springs inside like a crackling whip and clamps its jaws around one of Onta's tentacles before she can take cover behind her ink. Straightening its body, the eel begins to spin as though it were being turned on a spit, with each revolution twisting the octopus's arm until finally the limb is wrenched off. Now there is no safety for Onta in the strictness of her cave; like herself, the moray is at home in tiny spaces, and this first taste of her flesh will make it even more ferocious. Under these circumstances, to remain where she is means inviting being ripped apart and devoured piece by piece; however well she fights, she will probably be so badly mutilated she will die. Fortunately, there is an open seam in the rock at the rear of her cave, and while the moray is busy swallowing her severed tentacle, Onta presses her body through this slender crack and escapes.

Even if the eel had been able to pursue her, it would not have found a trail of blood to follow. As soon as Onta's arm was detached, the blood vessels exposed in the wound contracted and the flaps of skin ringing the broken flesh folded in to help seal the area. Several hours later, when she is more likely to be safe, the constricted vessels will expand again, allowing enough blood to flow into the wound for it to be capped with a protective clot. Within two days the wound will be completely healed, and then Onta's body will begin growing another arm to replace the one that was lost; a month or two from now, depending upon how much of the original arm is missing, she will have her new limb. This member may not be as long or as thick as the one taken by the eel, but it will serve her just as well. Given this ability to regenerate its tentacles, an octopus will readily sacrifice one in an act of self-mutilation if that arm has been pinned by an enemy and cannot be pulled free.

Alighting in what seems to be an uncrowded quarter of the reef, Onta barely has time to shield herself in the coral before she is again intruded upon. Springing toward her across the sand is one of her own kind, a beast of the same image but dangerous nonetheless. So great is the octopus's instinct for solitude that even things like itself are unwelcome inside the zone of its existence. Possibly Onta has landed too near the coral covering this octopus was using as a home, for the creature is bounding forward at a speed indicating that a fight is inevitable. At once, Onta's arms come alive and like lariats on the verge of being tossed they are twisting in tiny circles about her head.

Hercules and the Hydra

Poised this way, Onta is further testimony to the likelihood that the legendary Hydra was actually an octopus given to foraging on land near the port of Lerna in southern Greece. According to legend, this monster had a huge body and nine heads—eight that were ordinary and one that was immortal. From its lair in the Lernean marshes, the Hydra would crawl into the bordering fields and destroy the cattle there. Summoned to free the countryside from the creature's tyranny, Hercules forced it into battle by hurling flaming spears inside its cave. But each time he broke one of its mortal heads with his thundering club, the beast grew two others in its place. Only after he cut off its one eternal head did the Hydra die. This part of the corpse he buried in the earth, planting a heavy stone over it. Then he slashed open the wasted body and tipped his arrows with its poisonous gall.

As Onta loops her arms, they do indeed appear to be miniature heads surrounding the real one without which she would perish. If required to, the bottom-crawling octopus could steal across short stretches of land. In his observations of this creature (which are consistently accurate though made 2,300 years ago) Aristotle noted how it often was seen rising from the sea and walking over the coastal rocks. He also discussed its capacity for sprouting new arms and was aware of its venomous secretion. Writing in the age of the Roman Empire, Pliny has handed down tales of the octopus climbing onto harbor piers during the night, tempted by the barrels of salt fish waiting there to be hoisted aboard ship. One aquarium keeper of the last century actually discovered a captive octopus in a similar escapade. After dark the creature would ease out of its enclosure and slide along the wall like a mutant spider until it reached a tank containing some young lumpfish, dip into this ready supply of food for its evening meal, then tread back to its own pen. The excavation of Mycenae, a Peloponnesian city just a few miles from Lerna, has confirmed that the octopus was a familiar sight in the territory where the Hydra supposedly lived. Among the artifacts recovered there in 1876 were numerous golden ornaments hammered into the shape of an octopus with spiraling arms. Though we can never be entirely certain of what really took place in civilizations long gone, it does seem as if the Hydra was born in an unknown poet's imaginative rendering of an octopus.

As suddenly as it began its charge, the octopus rushing at Onta halts in front of her and raises its second and third arms, displaying the two disproportionately large suckers on each that only those of its sex possess. Confronted by the unmistakable sign of the invader's maleness, Onta also refrains from attacking, her behavior being willed now by an urge much stronger than her aversion to company. Fully grown but not yet old, she is ready to mate. In other species of the animal, the male reveals himself through the colors he bears as he moves forward, his body and arms being transformed into a brilliant parade of stripes on the side nearest his intended mate. Some require that he carry out his advances stealthily and with entwining arms catch the waiting female from behind. But in the grappling that follows there is no assault, no contesting and defeating—only the embrace of creation, the act of giving life.

At the moment of union, the one arm of the male capable of fulfilling this function slips into the female's body through the passage in her mantle. Down the length of this special limb, between the rows of suckers, runs a line of skin that is grooved; and it is along this narrow channel that packets of sperm flow from the male into his mate. Once inside the female, each pouch oozes a cementlike substance that fastens it into position near the base of the gills. (In a few species the sperm is encapsulated in the whitened end of the male's mating arm, and at the height of his excitement this potent tip breaks off and swims under its own motion into the female.) Some time later these tiny caches will burst, saturating the female with their seed, and then the mating is complete. But this may not happen until long after the two creatures have separated and are again wandering solitary ways.

In the spring of 1970, a young female octopus from the Pacific Ocean is delivered to the New York Aquarium, to be housed among its many exhibits. But without realizing it, her captors have done more than give the aquarium's visitors the chance to see for themselves the subtle motions of her life. Impregnated by a male be-

Grappling to mate

fore she was lifted from the sea, she soon begins to adorn the sides of her tank with the strands of eggs emerging from her body. It is a fabulous opportunity to study the earliest unfoldings of this creature's existence. For as long as the octopus has been observed, sufficient information has still not been gathered about the nature of its infancy.

While the babies are curled inside their capsules a few are made the subject of an experiment suggested by an old argument concerning the creature's ink. In this one it is believed that the octopus will perish if forced to remain in water heavily saturated with its own defensive cloud—a proposition advanced over the years by men who placed small adult octopuses in separate buckets and jars filled with seawater, provoked them into releasing their ink, and minutes later saw them die. Taking care not to harm the infant animals or pierce the membranes closed around them, the experimenters prod the babies until they have fired so much ink they can barely be seen through the translucent walls of their sacs. Nevertheless, they do not succumb. For as much as forty-eight hours they exist inside the blackness of their own mist; yet at the end of that time, they too are hatching normally.

Exactly how many days the eggs of an octopus must hang before the young are born varies from four weeks to four and a half months, depending upon the species and the temperature of its water. On August 19, the first baby octopuses to hatch are noticed in the tank. For the next two weeks they appear at the rate of a few every twenty-four hours, and for three weeks after that no more than thirty make it out in any one day. Then, in two consecutive days, thousands are hatched together; but then the swarming ends. Though the babies continue to be born for another three weeks, they never again emerge in such large numbers, the total for each day dropping back to the earlier levels.

As soon as the new arrivals are swimming freely they are transferred to a 50-gallon tank that has a circulation system designed to provide them with a steady surface current. While they are still healthy, the babies try as much as possible to stay in the upper perimeter of their pool, just as they would have headed upward to join the plankton had they been born in the sea. Not until they have lost their coordination and are swimming poorly do they settle among the rocks below. This time their preference for the bottom is not a sign that they are maturing, simply a sinking down that carries them closer to death.

Getting the infants to eat is the most immediate problem facing the tank men charged with keeping them alive. Initially they are offered brine shrimp that, like themselves, have just been hatched. Then finely chopped smelts are tried. Finally, fish eggs are dropped in their tank. These foods are not entirely ignored. Some of the babies grab at the pieces and even hold on to them as they swim about. But it cannot be said for certain that they are nibbling the things they have seized and are not merely playing with them. When a new method of feeding is tried and the meal is suspended in the tank on hooks, the baby octopuses are seen clinging to the individual particles but do not seem to be swallowing them.

Each day a few more die and are replaced in the tank by the fresh specimens hatching out. At different stages in their transition from embryo to emergent octopus the babies have been probed by the laboratory staff. On one occasion, several were removed from their eggs while still joined to the yolk and their subsequent development carefully watched. Though the whitish nutrient ball hanging from the mouth of each impeded its swimming, these infants stayed alive long enough to absorb what was left of the yolk and showed as much growth as those born at their natural time. Of particular interest to the scientists are the few babies that seem to have hatched with congenital defects. Since creatures born handicapped are easy prey and therefore not likely to be found in the ocean, the biologists have decided to devote most of their attention to these abnormal forms. Even in this special situation, however, they do not get many subjects. Of a thousand babies examined, only four are crippled in some way, though these do provide material that may never before have been investigated and give some indication of how many infants in a brood hatched at sea can be expected to emerge healthy.

By mid-autumn all the octopuses in the tank have died and no more are hatching. Had they been born in the ocean that once sheltered their mother, some of them would naturally have survived for two or three years—the full span of an octopus's life. Here their season was drastically curtailed—hardly begun. But it has not been wasted. Important information has been gained about the developmental stages of an animal whose earliest moments had not been adequately described. With this knowledge circulating among other scientists, the next baby octopuses born captive in a laboratory may be kept alive a while longer, enabling their custodians to discover something more about them. (At the University of Miami in Florida, a young graduate student, E. T. LaRoe, subsequently gathered from the Atlantic the eggs of two species of small tropical squid. For nearly five months the babies hatched in his tanks thrived on a diet of live mysids and larval fish, and when these cephalopods eventually died it was as adults, capable of spawning a new generation to succeed them.) By such tiny steps do the ocean sciences progress and people learn to see a creature they have feared in an enlightening way.

Alone in her new retreat, a thin tear in the rocks far enough from the reef to be infrequently visited, Onta tends to her developing babies, whose capsules dangle from the sheer walls like a curtain of beads. No longer will she range across the ocean bottom at night, beseiged by even the slightest rippling of the water nearby. At last her passion for isolation has been reinforced by a concern much greater in her than hunger. Even as her body starts to deteriorate from the persistent lack of food, she will not slacken her concentration on her young and leave them to hunt. By the time they are ready to hatch, it is obvious she is dying—too weak to save herself once they have gone. Unable to move, she watches them as they fill the cave, emerging like fireflies bejeweling a damp sky, then swim into the open, headed for the light—the beginning of their day.

EPILOGUE

*I*n May, 1879, Captain William H. Kerby of the schooner *Wm. V. Hutchins* was surprised by the appearance of a new species of fish at the end of his long lines. Looking for cod just south of Nantucket Lighthouse, he had hauled in a large, bluish-green fish splashed with yellow spots, its brilliant colors resembling those of a fish more commonly associated with tropical waters. Upon landing in Gloucester, he handed his discovery over to scientists who later confirmed that it was indeed a kind never seen before. Their subsequent investigations identified the creature as a bottom-dwelling fish that occupied a narrow band of water on the edge of the continental shelf that was continually warmed by the Gulf Stream. No sooner had the first tilefish been caught and named when others seemed to be turning up everywhere along the North American coast from Nova Scotia south to Virginia, though it was especially abundant off the southern New England shore. Like the wildflowers whose seeds lie buried in the desert ready to spring into bloom the instant there is ample rainfall, the tilefish had arrived in fabulous numbers, virtually overnight. Since it typically weighed between 15 and 25 pounds (with many individuals registering double that figure), made a hardy chowder and tasted good whether baked, broiled or fried, it was a welcome addition to the regular catch. In less than two years, an important fishery had grown up around it.

But early in 1882, a procession of heavy gales caused streams of unusually cold water to sweep the coast, and the Gulf Stream shifted further out to sea. Conditioned to living in the relatively shallow depths of the continental shelf, the tilefish may have been unable to follow the Gulf Stream out toward the Atlantic abyss, where the bottom drops off sharply. Moving onto the landward shoals would also have been hazardous as the water there was always unbearably cold. That spring, millions of tilefish were found floating on the surface, their gills turned red, their rose-colored bellies tilted toward the sun. For almost 4,000 square miles their corpses covered the ocean like stepping stones in a pond. As unexpectedly as it had appeared, the new species was gone—a victim not of pollution or human exploitation but of the constant caprice of nature, of change so swift and severe that it breeds catastrophe. Years passed without one tilefish being landed, and it was believed that the critical drop in the temperature of its portion of the sea had extinguished the animal for all time.

Yet as sensitive as this fish is to climatic variation, it was not entirely wiped out. Somewhere on the continental shelf, in caves burrowed in the soft sediment of the sea floor (which is the way the fish normally shelters itself), the hardier members of the species were resisting the cold flows that had invaded their home, still feeding on crustaceans and crabs and small fish. Like the guerilla remnants of a defeated army, the stronger tilefish were surviving out of sight, giving nature time to spin the wheel in their favor once more. Ten years after the holocaust, some tilefish were again seen in northern waters. By 1915, their recovery was so extensive that the fishing fleets of America and Canada were catching them in commercial quantities. Since then, the species has thrived without incident and more than 2,000,000 pounds of tilefish are being consumed each year.

Even for nature, which possesses an arsenal that includes monstrous earthquakes, temperature inversions, cosmic storms and atmospheric changes that can sour the air for a particular group of animals or plants, the eradication of stable organisms with a history of having successfully adapted to threatening conditions is exceedingly difficult. Once begun, life clings to itself with all the determination of a newborn baby tightening its fingers around any object placed in its palm and holding on with a strength that seems wholly out of proportion to its size. For every community of cells hopeful of withstanding the outer elements there must be a suitable shield, a container such as the human body, that houses it in a constant climate. Once this has been provided, survival is then largely a matter of how stable the circumstances to which the organism has molded itself remain. If they change, but gradually, by subtle degrees through many generations, then the creature in question has the greatest possible chance of evolving with them; of altering its structure to meet the progressive challenges of its environment.

Here is where life at sea has a critical advantage over all the plants and animals that have developed on land, except the insects. For the water itself is armor against the more violent upheavals of our planet, buffering and diffusing the consequences of geophysical events that are overwhelming on land and have in the past led to the sudden death of entire species there. Surely it is an imperfect armor, not without its own dangerous inconsistencies, as evidenced by the mass destruction of the tilefish; but ultimately it is one that offers each organism far greater protection than does a simple curtain of air. When life first came ashore, it may not have been in the form of an adventurous animal wilfully abandoning a deep, oxygen-rich ocean brimming with food. Whatever kind of creature it was that began the migration from sea to earth, it could have been acting out of desperation, coming from some clogged estuary, some decaying, draining pond that was choking its inhabitants. For those creatures born in more congenial waters, as most were, the sea was a nest from which there was no forced expulsion. Any animals discontented with conditions in the upper zones could also seek a more agreeable residence by guiding themselves downward; some eventually reaching the eternal blackness of the abyssal bottom. At this colossal depth, they dwell in almost perfect sanctuary, unaffected by wind and wave and tide; in frigid waters whose temperature is no different at the poles than it is at the equator. Here there is but one season, a climate held in perpetual suspension since its formation aeons ago when the ocean basins were filled. Devoid of vegetation, it belongs entirely to the solitary swimmers who bring untimely death to each other in the one place on Earth where all life is safe from the destructive surges of the planet itself.

To trace the genealogy of most ocean animals we must follow a path extending hundreds of millions of years into the past. Under the sea's protection, they have had an unheard-of amount of time to perfect their biological interplay with each other and the elements; to multiply their species, each bearing its own subtle refinement of the ancestral group's original plan for survival. Think of them as the small towns of the ocean existence, conditioned to live in long-settled ways, given room to modify themselves slowly. And the most successful of these roving communities are the mollusks, the phylum of invertebrates that reaches its zenith in the octopus, an animal whose capacity for thoughtful action prompted the late Loren Eiseley to comment on how fortunate human beings are that it never came ashore to stay. But some of its less intelligent relatives have. From inching snail to cave-like bivalve on up to the jetting cephalopod that might have rivaled us on dry earth, mollusks have advanced themselves through the sea, formed fresh water populations, and managed to colonize the land. One of its species, a garden slug, can even be said to inhabit the air as it mates while suspended above ground on filaments strung from any available bower. Compared with the history of this phylum, our own family tree is brief and inconclusive; an upstart sprout still struggling to clear the underbrush and bent on reaching the heights in a single day.

In the notes that accompanied the first paperback edition of his play, *Death of a Salesman*, Arthur Miller spoke of a Brooklyn that was more wilderness than city borough—a place of some dense woods, horses in pasture, and families that farmed vegetables in the fields surrounding their homes. A part of this century and the metropolis of New York, yet so unsettled that some of its citizens trekked miles in the spring mud to reach a store, coming from neighborhoods that had but a few houses and stood in their isolation like separate villages. That place of Miller's remembering no longer exists. But there is still something of a frontier in Brooklyn. Just across from the New York Aquarium, on the other side of the Coney Island boardwalk, is a slender stretch of beach; not the marsh it once was, but still enough of an unobstructed space to make a person feel free of city things. In summer it often seems to be just another urban street, packed with too many elbows, too much noise, paved with half-eaten sandwiches, soda cans and greasy wrappers. But once the weather has cooled and the crowds have gone and their debris has been buried by the swirling sand, the beach regains the beauty of an unencumbered shore. This is the season when a mantle of driftwood and seaweed rises on the damp, dark sand at the water's edge. Everywhere there is the sparkle of shells, and great flocks of gulls are no longer afraid to descend and sleep in open stand. Once again it is a reminder that not all the land is made of brick and steel and glass windows.

In the hearts of the children who venture here now, it is much more. Through its expanse, they see a savage land, desolate and forbidding, devouring all but the strongest, the noble. And its dragon is the wind, the angry, invisible roar that flies at their eyes and pins their arms to their sides as they butt forward, heading for the nearest jetty. Snatching up paper and small sticks as they go, the children race along the tidal flat to a place where the sea-crusted rocks form a hollow that they can huddle in and there they light their fire. First the ragged sheets of newspaper are twisted and rolled into rings and used to line the shallow pit that has been dug. From the store of fuel the youngsters have gathered they choose the thinnest and driest sticks, and these are laid on the paper so that they form a stack of wooden rectangles and the fire will rise through the spaces left between the pieces. Hunched around the one who holds the match, the children watch its magic flame being eased into each paper ring, feel safer being surrounded by its mounting cave of warmth.

Theirs is not the only blaze brightening the shore. Further up the beach there are small encampments of grown-ups; those few men and women who refuse to quit swimming even on days when the air has the numbing sting of dry ice. Some won't even be driven off by drizzle and mist. With their sheets of corrugated tin for blocking out the wind, their axes for splitting the stouter beams washed ashore, and the wire crates they use as cooking stands, they squat by the lap of the sea as though they meant to live there. They are all elderly, but it is difficult to fix a number to their age. They are so unlike the retirees who fill the boardwalk benches on clear days like a gallery of wrinkled statues, hoping a touch of sun will soothe their aches. Those people we can recognize as old within a prescribed range of years by their shuffling step, their uncertain grasp. It is evident in the way they cradle themselves with their gloved hands and crouch in their seats, their feet swaddled in blankets, as they close their eyes and slant their sallow faces to the sky. The beachmen no longer have smooth skin and they move in the measured pace of people who must husband their strength; but they do not seem to be deteriorating. It is not a desire for comfort that drives them; they have come for satisfaction, a feeling of being alive in the natural world. At a time when there is no shame in wanting to remain indoors, fully dressed, the sea draws them away from the city, urges them to swim.

One December afternoon when there was no wind and we felt brave enough to finish our lunch with a walk along the shore, we watched a man from their company curve through the swells with greater ease than we were strolling. Arching his back, he would slip under with barely a splash. As his legs disappeared from view, his head would emerge just long enough for him to swallow more air and then he would descend again. We never saw him move his arms. He must have extended them while he was submerged, gaining enough momentum with one powerful, sweeping stroke to surface and slide forward into his next dive. There was no break in his motion, no sudden jerking of limbs to impede his progress and define him as a human being trying to recall what it was like to be native to the sea. For as long as we watched him, he glided on as effortlessly as though he were a small whale. After a while, he no longer seemed completely human; we could only envision him as a vessel of mammalian intelligence, not wholly unlike the cetaceans he was emulating. Imagine, if you can, harpooning such a creature and processing it into soap, car wax, cosmetics, lubricants, luggage, margarine, fertilizer and pet food, as we routinely do the great whales. Now try to explain why, when there are adequate substitutes at reasonable prices for all these products.

Like the beachman who finds his harmony in the supple undulations of the aquatic mammal, we are all exiles from the sea, compelled to reach back to it for some element that will help sustain us. Whether the missing ingredient in our existence be biological knowledge that is translatable into life-restoring medicines, a continuing supply of food and minerals, or merely the taste of adventure that camping beside giant waters provides, the sea gratifies our senses in an irreplaceable way. As alien and inhospitable as it sometimes seems to us now, it is still home.

GLOSSARY

arrow-worm. Part of marine plankton belonging to the phylum Chaetognatha (Gk *chaite*, "hair," and *gnathos*, "jaw"). Their importance in the ecology of marine plankton far outweighs the number of species in the phylum. The arrow-worm has a transparent, torpedo-like body. It feeds upon other members of the zooplankton. Arrow-worms are hermaphroditic. Although there are only about sixty species they can dominate a water mass by the large numbers of individuals. Arrow-worms are important in the ecology of marine plankton feeding upon eggs and young stages of some of the commercially important fishes. Species of arrow-worms are so sensitive to even minor changes in the physical and chemical characteristics of water masses that they are used as biological indicators. When one species of arrow-worm (*Sagitta elegans*, L *sagitta*, "arrow," and *elegans*, "select") replaces another

(*Sagitta setosa*, L *sagitta*, "arrow," and *setosus*, "bristly") in the North Sea during summer and early fall it signifies an influx of water from the northern Atlantic. This water becomes richer in micronutrients and presages greater productivity, especially with regard to herring fishery.

arthropod. The phylum Arthropoda (Gk *arthron*, "joint," and *pous*, "foot") consists of more species (over 800,000) than all other animal phyla added together. They are the only invertebrates to have succeeded greatly in adapting to life on land and the only animal group other than the vertebrates to adapt to flight. The arthropods consist of three large groups: the insects (more than 700,000 species); the arachnids (spiders, scorpions, mites, ticks, etc., more than 47,000 species) and the crustacea (about 30,000 species) and a number of minor groups. The insects and arachnids are primarily land animals; the crustacea reside mostly in the marine environment but have many representatives in fresh water and a few on land. The arthropod has an externally segmented and jointed body with a chitinous exoskeleton divided into head, thorax and abdomen, and paired jointed appendages. These animals grow by molting, i.e., shedding their exoskeleton.

baleen. (also known as whalebone) The fibrous plates attached to the upper jaw of certain whales (blue, fin, right, gray, etc.). It is used to filter food organisms (e.g., krill) from sea water.

bioluminescence. The emission of visible light by living organisms. Often referred to as "cold light," it is the same as that produced on land by fireflies. Many marine creatures are bioluminescent, including bacteria, one-celled organisms, shrimps, squid, various species of fish. Luminescent organisms are more numerous in the marine environment than on land or fresh water.

buccal membrane. The lining of the buccal cavity (an area primarily concerned with feeding) which sometimes serves as a point of attachment for spermatophores in certain species of squid.

cetacean. A member of the aquatic mammals consisting of whales, dolphins and porpoises. Cetacea are divided into the Mysticeti whales (Gk *mystax*, "mustache," and *ketos*, "whale": whalebone or baleen) and Odontoceti (Gk *odontos*, "tooth," and *ketos*, "whale," toothed whales: dolphins, narwhal, beluga, sperm whale, etc.).

chromatophore (Gk *chromatos*, "color," and *phero*, "bear or carry"). A pigment-bearing cell found in the integument of various animals (e.g., squid) capable of changing the color of the skin by contracting or expanding.

cilia (L *cilium*, "eyelash"). Permanent and highly vibratory ectoplasmic structure of certain cells. They are found in almost all groups of the animal kingdom fulfilling varied and important functions, such as locomotion and the passage of food particles.

copepod (Gk *kope*, "oar," and *pous*, "foot"). Small crustaceans abundant in both fresh and salt waters. Most are free-swimming but some are parasitic on the skin and gills of fish. Free-living copepods are important in the food chain in the sea.

crustacean (L *crusta*, "shell"). Important class of the phylum Arthropoda. Crustacea comprise crabs, lobsters, shrimps, barnacles, copepods, etc. Many act as scavengers and are themselves the food for larger marine animals. Others live on diatoms and other minute organisms and provide food for fishes and even the gigantic whales (e.g., krill eaten by blue whale).

ctenophore (Gk *ktenos*, "comb," and *phero*, "to bear"). Commonly called comb jellies or sea-gooseberries. There are about eighty species, all aquatic. They are transparent with eight vertical rows of plates of fused cilia (the *ctenes* or comb plates). Since the *ctenes* are iridescent, many ctenophores swimming in still water with the sunlight bouncing off them form a beautiful sight.

diatom. Unicellular or colonial algae having silicified cell wall that persists as a skeleton after death and forms diatomite (also called diatomaceous earth which is used as a filtering material, as an absorbent, filler for paints and plastics, and as an abrasive). Diatoms frequently constitute the bulk of marine and fresh water phytoplankton.

dinoflagellate (Gk *dinos*, "whirling," L *flagellum*, "whip"). Unicellular algae that whirl about by means of flagella. Some species are of economic importance because they are responsible for the phenomenon known as "red tide."

echiuroid (Gk *echis*, "adder, viper" and *oura*, "tail"). All eighty species are marine inhabitants. This animal has a body that is divided into a long proboscis, used to gather food, and a bulbarlike trunk. *Bonellia viridis* exhibits sexual dimorphism, the males being minute, living within the females.

epiphyte (Gk *epi*, "on" and *phyton*, "plant"). A plant living upon another plant but not parasitically. It obtains its nutrients from the surrounding environment.

larva (L *larva*, "ghost, mask"). An independent, immature form of an animal which is unlike its parents and must undergo some sort of metamorphosis before assuming the characteristics of the adult.

metabolite (Gk *metabolos*, "changeable"). A product (foodstuff, intermediate waste product, etc.) of metabolism by which small molecules result from large molecules and vice versa.

metamorphosis (Gk *meta*, "change" and *morphe*, "shape or form"). A normal process in the life history of many animals which results in a radical change of structure within a relatively short time. For example, a sea urchin larva undergoes metamorphosis into the adult.

mollusk (L *molluscus*, "soft"). An animal phylum consisting of over 100,000 species. Most are aquatic, both marine and fresh water, but there are some land representatives (e.g., garden slug). The phylum includes snails, octopuses, squid, clams, mussels, oysters, etc.

morphology (Gk *morphe*, "shape or form" and *lego*, "choose, speak"). The study of the configuration or structure of animals and plants.

pangea (Gk *pan*, "all, the whole" and *ge*, "earth"). The postulated former supercontinent composed of all the continental crust of the earth and later fragmented into Laurasia and Gondwana.

phosphorescence. Term often used for the brilliant displays of bioluminescence in the sea (cf. bioluminescence). One of the bays in Puerto Rico is so spectacularly alive with bioluminescent organisms that it is called Phosphorescent Bay (Bahia Fosforescente).

phytoplankton (Gk *phyto*, "plant," and *planktos*, "wandering"). Plants that drift and abound in the sea, brackish and fresh water. All sea animal life is directly or indirectly dependent on phytoplankton productivity.

pteropod (Gk *pteron*, "feather or wing," and *pous*, "foot"). Commonly called the sea butterflies. These snails with or without a shell swim at the surface of the sea by means of a pair of lateral pedal extensions reminiscent of butterfly wings.

sea hare. A gastropod mollusk in which the shell is reduced and internal, unlike the more typically shelled snails.

siphonophore (Gk *siphon*, "tube," and *phero*, "bear"). A colonial hydrozoan whose most infamous representative is the Portuguese man-of-war. It is, therefore, not a true jellyfish. Jellyfish move by actively pulsating motions; the Portuguese man-of-war with its purplish gas bag is moved about by winds and currents.

spermatophore (Gk *sperma*, "seed," and *phero*, "bear"). A bundle or packet of sperm produced by male members of certain animals such as the squid, octopus, and some vertebrates.

tintinnid (probably from L *tintinnabulum*, "bell"). This is a conical or bell-shaped unicellular ciliated organism. It has an outer shell known as the lorica (Latin for corselet).

zooplankton (Gk *zoön*, "animal," and *planktos*, "wandering"). Usually the eggs and young stages (larvae) of many animals are part of the "drifting" animals in the sea. Many are important constituents of the food chain.

INDEX

Acetylcholinesterase, 44
Adeinine arabinoside, 117
Aequorea aequorea, 97
Aequorin, 97
Algae, 17, 29, 30, 32, 33, 74, 118, 119
  green, 13
Alginates, 33
Alginic acid, 33
Anchovy, 27
Anemone, *see* Sea anemone
Angelfish, 133
Anguilla anguilla, 42
Anguilla rostrata, 43
Antibiotics, 23, 24, 118
Antibodies, 124, 125
Ara-A, 118
Ara-C, 116–117
Aristotle, 140
Arrowworms, 26, 27
Asahi Chemical Industries, 117

Barnacle, 57
Bergmann, Werner, 116–117, 118
Bigeyes, 108
Blenny, 100

Bonellinin, 90, 92
Butterfly fish, 100, 107

Calcium alginate, 33
Cardinalfish, 107
Carrageenin, 34, 38, 40
Cephalosporins, 24
Chimpanzees, 72
Cholesterol, 33
Chondrus crispus, 34
Clams, 21, 110, 112, 119
Conger eel, 136
Copepods, 24
Coral, 9–12, 73–78, 94–97, 100, 107–109, 114, 119
  brain, 107
  gorgonian, 95–96
Crabs, 109, 113–114, 119, 122, 127, 132
  pea, 119
Crown-of-thorns, 73–78
Crustaceans, 26, 27, 108
Ctenophores, 27
Curare, 21
Cytosine arabinoside, 117

Damselfish, 107
Diatoms, 17, 18, 61
Dinoflagellates, 17, 18, 20–21, 30–31, 113
Diseases, oceans' creatures and, 122–125
Diving-reflex phenomenon, 83–86
Dolphins, 65, 81
Drugs from the sea, development of, 20–24, 92–94, 116–118

Ebimar, 41
Echinoderms, 97–98, 100
Eel
  American, 43, 45
  conger, 136
  electric, 43–44
  European, 42–43, 45
  moray, 132–133, 136–137
Eiseley, Loren, 156
Eledoisin, 131
Epiphytes, 40

Flounders, yellowtail, 123
Food and Drug Administration, 118

Giordano, Steven, 38
Gobies, 107
Gold, Kenneth, 31
Great Barrier Reef, 74, 76
Grunts, 108
Gulf Stream, 152

Hapalocheana maculosa, 130
Hardy, Sir Alister, 2E
Hercules, 140
Herring, 29, 123
Holothurin, 98
Hugo, Victor, quoted on octopus, 128
Hunt, James P., 38
Hydra, 140, 141

IgM protein, 124–125
Immunoglobulins, 124
International Whaling Commission, 63, 70
Irish moss, 34, 36–40
Ishida, Torao, 117

Jellyfish, 26, 27, 49, 96–97, 112

Kainic acid, 32
Kerby, William H., 151
Kobayashi, Hidehiko, 117
Krill, 29, 61, 62
LaRoe, E. T., 147
Laurencia pacifica, 30

Lilly, John C., 81
Lobsters, 109

Medical research, 20–24, 44, 48, 92–94, 98, 116–118
Miller, Arthur, 156–157
Moeller, Henry W., 38–40
Mollusks, 13, 21, 39, 60, 76, 108, 110, 112, 156
Moray eel, 132–133, 136–137
Moss, Irish, 34, 36–40
Mussels, 36, 39
Mysticeti, 61

Nereistoxin, 45
Nerve conduction, 47–48
New York Aquarium, 142
New York Ocean Science Laboratory, 38
Nigrelli, Ross F., 97–98, 123
Nitta, S., 45
Nucleosides, 116–117

Octopus, 13–14, 16–17, 24, 26–27, 42, 46, 49, 57–60, 82, 90, 94–97, 100–101, 106, 109–110, 112–114, 119, 120, 122, 126–133, 136–142, 144–148, 156
Oikopleura, 30
Orcas (killer whales), 64–66, 80
Organophosphates, 44–45
Osborn Laboratories, 90, 97, 98, 123
Oysters, 13, 21, 110

Padan, 45
Parke-Davis & Company, 117
Parrotfish, 107, 133
Payne, Roger, 81
Pearlfish, 97
Penguins, 29, 49, 64
Penicillin, 24
Pharmacology, marine, 20–24, 92–94, 98, 116–117
Photosynthesis, 17
Phytoplankton, 17–18, 29
Plankton, 17–18, 20, 26, 30, 32, 60, 145
Pliny, 140
Prostaglandins, 96
Proteins, 124–125
Protozoa, 27, 30
Pteropods, 27, 60

"Red tide," 20, 24
Regeneration, 97–98, 100, 130
Research, medical, 20–24, 44, 48, 92–94, 98, 116–118

Sargasso Sea, 42–43
Sargassum weed, 42
Saxitoxin, 20–21
Scallops, 112
Sea anemone, 13, 127
Sea cucumber, 97–98
Sea fans, 12, 119
Sea hare, California, 30
Sea squirts, 13, 92–93
Sea stars, 73–78, 98, 113
Sea whips, 13
Sea worms, 90
Seals, 49, 64, 83
Seaweeds, 22, 32, 33, 34, 36–40, 108
Sharks, 49, 50, 60, 120, 122, 124–126
Shock, 83–86
Shrimps, 109, 122
Siegel, M. Michael, 93, 124
Siphonophores, 27
Slug, garden, 156
Snails, 13, 60, 110, 114, 127
Soapfish, 108
Sponges, 13, 113–114, 116–119
Squid, 45–54, 88–89, 147

Squirrelfish, 108
Staphylococci, 24
Stingray, 120, 122
Strontium, 33
Surgeonfish, 107

Tang, 133
Tethya crypta, 116
Tilefish, 152–153, 154
Tintinnids, 31–32
Tritons, 76
Turtle weed, 113, 119

Walrus, 65
Weinheimer, Alfred J., 92, 96
Whales, 28, 80, 83–85, 88, 160
  blue, 61–66, 70–71
  humpback, 66, 81
  killer, *see* Orcas
  sperm, 49, 82–84, 86, 88–90
  toothless, 61
Williams, Leonard W., 48
Wool, alginate, 33
Worms, 112
Wrasses, 106, 126–127

Young, J. Z., 48

Zooplankton, 24, 27–29, 61, 65